Universitext

Universitext

Universitext is a series of textbooks that presents material from a wide variety of mathematical disciplines at master's level and beyond. The books, often well class-tested by their author, may have an informal, personal even experimental approach to their subject matter. Some of the most successful and established books in the series have evolved through several editions, always following the evolution of teaching curricula, to very polished texts.

Thus as research topics trickle down into graduate-level teaching, first textbooks written for new, cutting-edge courses may make their way into *Universitext*.

For further volumes:
http://www.springer.com/series/223

Pablo Amster

Topological Methods in the Study of Boundary Value Problems

 Springer

Pablo Amster
Departamento de Matemática
FCEN-Universidad de Buenos Aires
    and IMAS-CONICET
Buenos Aires, Argentina

ISSN 0172-5939   ISSN 2191-6675 (electronic)
ISBN 978-1-4614-8892-7   ISBN 978-1-4614-8893-4 (eBook)
DOI 10.1007/978-1-4614-8893-4
Springer New York Heidelberg Dordrecht London

Library of Congress Control Number: 2013948451

Mathematics Subject Classification: 34B15, 34C25, 37C25, 35J25, 47H10, 47H11, 47J25, 46T99

Printed on acid-free paper

Springer is part of Springer Science+Business Media (www.springer.com)

Preface

This book constitutes an elementary introduction to the application of topological techniques in nonlinear analysis. For the reader's convenience, only boundary value problems for ordinary differential equations are treated, although most of the ideas can be generalized for partial differential equations and some other areas of mathematics. This approach will allow the student to avoid many of the technical difficulties related to nonlinear problems and focus on the application of topological methods. Only basic knowledge of the main topics in analysis is needed, making the book easy to understand for nonspecialists, too. Whenever possible, just elementary tools are employed; in particular, in many situations we arrive at important and nontrivial results by means of elementary techniques. Despite its simplicity, the main ingredients of nonlinear analysis are present, so readers with some experience in functional analysis or differential equations may also find some elements that complement and enrich their tools for solving nonlinear problems in many different fields. The style throughout the book is concise and informal, allowing students of all levels to have a first glimpse at this interesting and beautiful branch of mathematics.

This book could never have existed without the support of my colleagues, students, and friends (most of them belong to at least two of the listed categories): Rafael Ortega, Mónica Clapp, Colin Rogers, Man Kam Kwong, Alfonso Castro, Lev Idels, Jorge Cossio, Julián Haddad, Pablo De Nápoli, Juan Pablo Pinasco, Diego Rial, Paula Kuna, Alberto Déboli, Manuel Maurette, and Rocío Balderrama. All that I've learned about mathematics is due to them. Also, I'm grateful for the support of my family and the team of "nonmathematical" friends from whom I've learned all that I know about life. I want to give special thanks to Donna Chernyk for trusting in this project and for all her help and patience and to the anonymous reviewers for all the corrections and comments that helped to improve the first version of this manuscript. All remaining mistakes and flaws of the text are my sole responsibility.

Buenos Aires, Argentina   Pablo Amster

Introduction

Nonlinear analysis is a field with a large number of applications in various sciences. In particular, the study of boundary value problems for differential equations has been the subject of intense research in recent decades. Many different techniques have been developed for the study of nonlinear problems; among them, one of the most effective techniques consists in the use of diverse topological methods, such as the shooting method, fixed point theorems, upper and lower solutions, or degree theory. In particular, topological methods have proved to be successful for many problems that have no variational structure. A first basic result in the direction of the fixed point methods is the now well-known Banach fixed point theorem, which generalizes the method of successive approximations proposed by Picard. Despite its simplicity (and the fact that it is almost 100 years old), the Banach theorem is still an efficient and popular tool for proving many different existence-uniqueness results. However, in some cases the conditions of this theorem are too restrictive, and more powerful techniques are required. An example is the Schauder fixed point theorem, which can be regarded as an extension, for a compact operator in a Banach space, of the well-known Brouwer theorem and is especially useful in the field of differential equations, where the associated operators usually have a compact inverse in some appropriate space. The Schauder theorem can also be used to develop the method of upper and lower solutions, which makes it possible to prove the existence and location of solutions for a very general class of equations. Also, there are some fixed point theorems in cones of the Krasnoselskii type that have applications to some specific problems, for instance, many proofs of existence of positive solutions of some differential equations, frequently found in real-world applications, are based on these kinds of results.

All the aforementioned techniques can be introduced within the more general scope of topological degree theory, which, roughly speaking, is an algebraic count of the zeros of a continuous function defined over a bounded subset of a normed space. In the finite-dimensional case, it was defined by Brouwer and provides a straightforward proof of his fixed point theorem, among other results. The extension to a general Banach space is due to Leray and Schauder and assumes that the function is a compact perturbation of the identity, namely, an operator of the type $I - K$, with K compact. Again, this setting can be regarded as "natural" in the context of many boundary value problems for differential equations. Besides the solution property, which ensures that a mapping with nonzero degree has a zero, one of the most powerful properties of the degree is its homotopy invariance that makes it possible, under appropriate conditions, to transform a problem into a simpler one for which the degree is easy to compute. The equation under study is thus embedded into a one-parameter family of equations; the existence of a priori bounds of the solutions guarantees that the degree will be constant over the deformation. The topological degree is particularly useful in so-called *resonant problems*, those in which the associated linear operator is noninvertible, and hence it is not always clear how to convert them into a fixed point equation.

The book is self-contained, in the sense that only basic notions of analysis are needed to understand most of the contents. The examples mainly concern boundary value problems for ordinary differential equations. In most cases, we shall take as a model equation the second-order problem

$$u''(t) = f(t, u(t), u'(t)), \qquad 0 < t < 1,$$

with f continuous, under the boundary conditions

$$u(0) = u(1) = 0 \qquad \text{(Dirichlet)},$$

$$u'(0) = u'(1) = 0 \qquad \text{(Neumann)},$$

or

$$u(0) = u(1), \qquad u'(0) = u'(1) \qquad \text{(periodic)},$$

among others. The latter conditions can be interpreted as truly periodic when $f(t+1, u, v) = f(t, u, v)$ for all u, v, and $t \in \mathbb{R}$: in this case, a solution $u : [0,1] \to \mathbb{R}$ can be extended periodically to a C^2 function defined in the whole real line. This setting will be particularly useful when dealing with some *delay differential equations*, in which the nonlinear term f also depends on $u(t - \tau)$ for some $\tau > 0$. In some cases, our model equation will in fact be a *system* of n equations: as we shall see, this extension is not always trivial and may involve some geometrical or topological difficulties.

The results presented here are not the best of all possible results, in the sense that, in most cases, we prioritized giving an intuitive and easy approach over obtaining better or sharper theorems. We focus all the time on the methods and on the specific problems; in particular, this is one of the reasons for which all examples refer only to ordinary differential equations, which makes it possible to avoid some technicalities. Also, we have chosen to work always in the spaces of continuous or continuously differentiable functions, although better results can be obtained using, for example, Sobolev spaces. In many cases, the same problem is studied using different tools, so it may happen sometimes that a result presented in one chapter is improved in a later one. The reader may also feel that some of the computations required for the different methods are unnecessarily repeated in different chapters, but this was done for the sole purpose of preserving the "self-containedness" spirit of the text.

The book is organized as follows. In Chap. 1, we introduce one of the simplest topological methods, usually known as the *shooting method*, which basically consists in reducing a problem to a finite-dimensional equation for a certain parameter λ. Then, appropriate tools can be used, such as the Brouwer fixed point theorem or equivalent results. The chapter was designed to be self-contained and employs only concepts from basic calculus; for simplicity, our study of systems is restricted here to the two-dimensional case, for which we present a very elementary proof of the fixed point theorems we shall be using. There are many extensions and improvements of the basic results, which require slightly more advanced topics (e.g., Stone–Weierstrass theorems); for this reason, they were included in starred sections. This does not mean, of course, that they are extremely difficult: the idea was just to show that most of the topics—the nonstarred sections—could be understood within the context of a first course in calculus.

The next chapter is devoted to the Banach fixed point theorem and some of its immediate consequences. In particular, we shall prove the usual version of the implicit function theorem in Banach spaces and present some applications to boundary value problems. This requires a knowledge of the basic notions of differentiation in Banach spaces, which for the sake of completeness are presented here.

In Chap. 3 we develop some iterative methods such as the monotone iterations method and the Newton method and some of its variants. Applications are given to some boundary value problems. Also, we introduce a Cantor diagonalization argument, which makes it possible to deal with problems in unbounded domains.

In Chap. 4 we prove the general version of the Brouwer theorem and a well-known extension to Banach spaces: the Schauder theorem. Among other uses, this latter result allows us to give a complete version of the method of upper and lower solutions introduced in the previous chapter, with applications to many different problems. As a corollary, we obtain the so-called Schaefer theorem, which can be regarded as a particular case of the Leray–Schauder continuation techniques.

These techniques require a more sophisticated topological tool: the aforementioned topological degree, constructed with some detail in Chap. 5. As usual, the Brouwer degree is introduced first and then extended for compact perturbations of the identity in a Banach space, namely, the Leray–Schauder degree. The specific difficulties of the construction are not essential for the applications, so readers not particularly interested in certain topological issues may avoid most of the contents in this chapter and keep in mind only the main properties of the degree mapping.

Finally, in Chap. 6 we present applications to various boundary value problems. Starting with specific examples, we obtain some general continuation theorems that can be applied in many situations. In particular, most of the sections are devoted to the study of resonant problems, for which we discuss some classical results and different extensions.

For the reader's convenience, we include a brief review of the main results from the theory of ordinary differential equations used in the book. The list reduces to some fundamental theorems such as existence and uniqueness, continuous dependence, and a few specific facts concerning second-order equations. Also, we give an account of some definitions and elementary aspects of the general theory of metric spaces. Some specific results used in various chapters are commented upon in the respective appendices.

Incidentally, a few basic facts of measure theory shall be mentioned or employed in some particular examples, such as the Lebesgue measure of a set or some well-known convergence theorems like the dominated convergence theorem or the Fatou lemma. In a (unique) specific exercise we refer to convolution and mollifiers, but all of this can be completely skipped by the uninterested reader.

All chapters end with a list of exercises. For convenience, there is a section at the end of the book with hints and solutions to selected problems. Every exercise, no matter how easy, deserved at least one or two lines in this latter section: sometimes just a short comment might be of some help or provide a different point of view. Complete resolution is not the general rule, although some more difficult problems are solved in a reasonably detailed manner.

To conclude this introduction, it is worth mentioning that the bibliography at the end of the book includes not only the various papers and books cited in the different chapters but also a rapid account of general texts that cover most of the topics addressed in the text e.g. [3, 4, 22, 26, 32, 81, 93, 104, 107]. The list is not exhaustive; it was intended to provide a basic overview of the various paths the reader might take to tackle more in-depth studies on the subject.

Contents

Notation and Special Symbols

Throughout the book, we shall use the standard terminology for sets and relations. In particular, we shall use $\mathbb{N}$, $\mathbb{Z}$, $\mathbb{R}$, and $\mathbb{C}$ for natural, whole, real, and complex numbers, respectively. Also, we shall use $'$ for derivatives of a real function and $\frac{\partial}{\partial x_j}$ for partial derivatives. The notation "D" is reserved for the differential of a mapping. Other typical symbols and notations shall appear, such as, for example, lim, lim sup, and lim inf. The closure of a subset A of a metric space X shall be denoted by $\overline{A}$. The distance between two elements $x, y \in X$ shall be denoted by $d(x,y)$. An open ball of radius r centered at a point $x \in X$ shall be denoted by $B_r(x)$. In a few specific situations we shall employ the distance between a point and a subset $A \subset X$, denoted by $dist(x,A)$. If L is a linear operator between two vector spaces, we shall use the standard notation $Ker(L)$ and $Im(L)$ to refer to the kernel and range or image of L, respectively. In the case of a matrix $A \in \mathbb{R}^{m \times n}$, we shall write a_{ij} for its ij entry. The isomorphism between $\mathbb{R}^{m \times n}$ and the space $L(\mathbb{R}^n, \mathbb{R}^m)$ of linear transforms from $\mathbb{R}^n$ to $\mathbb{R}^m$ shall be ignored: in other words, for $x \in \mathbb{R}^n$, we shall avoid the use of transpose symbols and write directly $Ax \in \mathbb{R}^m$. When $n = m$, the determinant of A shall be denoted by $det A$.

Unless specified, when X is a normed space, the norm of an element $x \in X$ shall be written, as usual, $\|x\|$. Special cases include the following:

- $\mathbb{R}^n$, for which we shall use the notation $|\cdot|$;
- The space $C([a,b])$ of continuous vector functions $u : [a,b] \to \mathbb{R}^n$, with norm

$$\|u\|_\infty := \max_{t \in [a,b]} |u(t)|.$$

More generally, for $k > 0$ we shall denote by $C^k([a,b])$ the Banach space of C^k functions from $[a,b]$ to $\mathbb{R}^n$, equipped with the norm

$$\|u\|_{C^k} := \max\{\|u\|_\infty, \|u'\|_\infty, \ldots, \|u^{(k)}\|_\infty\}.$$

Occasionally, we shall employ the space C_T of continuous T-periodic functions $u : \mathbb{R} \to \mathbb{R}^n$, also equipped with a uniform norm;

- The usual Lebesgue space $L^2(a,b)$ of measurable vector functions $f : (a,b) \to \mathbb{R}^n$ such that

$$\|f\|_{L^2} := \left(\int_a^b |f|^2 \right)^{1/2} < \infty.$$

In general, an inner product defined over a vector space shall be denoted by $\langle \cdot, \cdot \rangle$. This includes the case of $L^2(a,b)$, with

$$\langle f, g \rangle := \int_a^b f \cdot g.$$

For the euclidean inner product of two elements $x, y \in \mathbb{R}^n$ we shall directly use the dot notation $x \cdot y$, that is,

$$x \cdot y := \sum_{j=1}^n x_j y_j.$$

The maximum, minimum, and average of a continuous function $u : [a,b] \to \mathbb{R}$ shall be denoted respectively by u_{max}, u_{min}, and $\bar{u}$, namely,

$$u_{max} := \max_{t \in [a,b]} u(t), \qquad u_{min} := \min_{t \in [a,b]} u(t), \qquad \bar{u} := \frac{1}{b-a} \int_a^b u(t)\, dt.$$

When $u : [a,b] \to \mathbb{R}^n$, we shall use the same notation $\bar{u}$ to denote its average coordinate by coordinate, that is,

$$\bar{u} := (\bar{u}_1, \ldots, \bar{u}_n).$$

Some specific terminology and symbols such as *deg* for the degree of a function, *co* for the convex hull of a set, or *meas* for measure shall be defined within the particular context in which it is used.

Chapter 1
Shooting Type Methods

1.1 Set the Angle and Shoot

In this chapter, we shall describe a very elementary tool for the study of boundary value problems, usually known as the *shooting method*. In very general terms, the method can be summarized in two steps:

1. Solve an initial value problem with a free parameter λ.
2. Find an appropriate value of λ such that the obtained solution satisfies the desired boundary condition.

The task looks really simple, but it requires some qualitative analysis on the behavior of the solutions of the initial value problem, according to the variations in the parameter λ. As we shall see, in many cases it is possible to obtain enough information in advance to guarantee the success of the method.

Let us start with the second-order equation

$$u''(t) = f(t, u(t)) \tag{1.1}$$

with homogeneous Dirichlet conditions

$$u(0) = u(1) = 0. \tag{1.2}$$

Assume we know little about boundary value problems at this point, so we should start doing what we undoubtedly know: solve, in the first place, on Eq. (1.1) with initial conditions

$$u(0) = 0, \qquad u'(0) = \lambda \tag{1.3}$$

for fixed $\lambda \in \mathbb{R}$. Then, hopefully, we will be able to find a value of the shooting parameter λ such that the corresponding solution will also satisfy the boundary condition at $t = 1$.

To do things properly, assume that f is continuous and locally Lipschitz in u; then the solution u_λ of (1.1)–(1.3) is well defined and unique. Thus, it suffices to find λ such that $u_\lambda(1) = 0$; in other words, we are looking for a zero of the function ϕ defined by

$$\phi(\lambda) := u_\lambda(1).$$

P. Amster, *Topological Methods in the Study of Boundary Value Problems*, Universitext, DOI 10.1007/978-1-4614-8893-4_1, © Springer Science+Business Media New York 2014

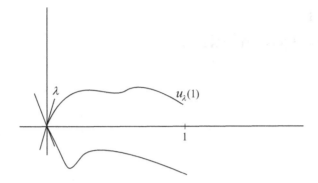

This explains the title of this section: indeed, the procedure simply represents an attempt to adjust the value of the parameter λ until an appropriate shooting angle is obtained, so the graph of u_λ hits the point $(1,0)$. This recalls somewhat an old computer game called *Gorilla*, now replaced by the very popular (and slightly more sophisticated) *Angry Birds*.

But there is always a *but*: observe that the solutions of the initial value problem may not be defined up to $t = 1$, and hence ϕ is not necessarily defined for all values of λ. However, we do know at least one very important fact about ϕ: it is continuous over its domain, no matter what this domain looks like.

In some lucky situations, it is possible to assert that $dom(\phi) = \mathbb{R}$; for example, this is the case when f grows at most linearly in its second variable, that is,

$$|f(t,u)| \leq a|u| + b$$

for some constants a and b (see Appendix B.1, Exercise 4 for details). In particular, if f is bounded, then the shooting method works perfectly well, as we shall see in the next section.

1.1.1 A Back-and-Forth Example: The Pendulum Equation

Let us illustrate the technique described previously with a well-known problem, the pendulum equation

$$u''(t) + \sin u(t) = p(t), \qquad 0 < t < 1,$$

where the forcing term $p : [0,1] \to \mathbb{R}$ is continuous. This equation is very famous; it has been the subject of many relevant works and is still being studied: for instance, the problem of finding all possible forcing terms such that periodic solutions exist has not yet been solved completely (for an account of the history and open problems on the pendulum equation, see, for example, the excellent survey [79]. More shall be said about the subject later on, in Chap. 3). However, the situation is completely different when dealing with the Dirichlet conditions (1.2), for which the existence of solutions can be easily proven by the shooting method.

Indeed, if u_λ is the unique solution satisfying the initial value conditions (1.3), then integration yields

$$u_\lambda'(t) = \lambda + \int_0^t [p(s) - \sin u_\lambda(s)]\, ds$$

and, hence,

$$|u_\lambda'(t) - \lambda| \le \int_0^t |p(s) - \sin u_\lambda(s)|\, ds \le \int_0^1 |p(t)|\, dt + 1.$$

Setting $R := \int_0^1 |p(t)|\, dt + 1$ we deduce that u_λ is monotone for $|\lambda| \ge R$, more precisely:

- $\lambda \ge R \Rightarrow u_\lambda'(t) \ge 0$ for all t,
- $\lambda \le -R \Rightarrow u_\lambda'(t) \le 0$ for all t.

In particular, u_R is nondecreasing and u_{-R} is nonincreasing; together with the fact that $u_\lambda(0) = 0$ and $u_\lambda'(0) = \lambda$, this implies

$$\phi(R) > 0 > \phi(-R).$$

Thus, by Bolzano's theorem we conclude that ϕ vanishes in $(-R, R)$.

The same procedure can be applied to Eq. (1.1) for arbitrary bounded f. But the boundedness condition is too restrictive: if we are willing to say that the shooting method is a useful tool for many different problems, then it would be desirable to see it applied in more general cases. This is the goal of the next section.

1.1.2 A Priori Bounds

From the previous example we may conclude, as a general rule, that the success of the shooting method relies very strongly on the fact that we know some properties of the associated flow of the differential equation. For instance, we need to be sure that the solutions of the initial value problem for an appropriate set of parameters are defined up to the endpoint of the interval. As mentioned, from standard results in the theory of ordinary differential equations, this is guaranteed for all λ when the nonlinearity has linear growth; however, there are plenty of cases in which this restriction is not satisfied and the shooting method is still applicable. In particular, in some situations it might be very helpful to be equipped with count with a priori bounds of the solutions.

The philosophy behind this idea is again very simple: if we know in advance that the solutions of a certain problem are bounded by some constant R, then we may replace the nonlinearity f by a bounded one, say $\tilde{f}$, such that $\tilde{f}(t,u) = f(t,u)$ for $|u| \leq R$. Obviously, this must be done in such a way that the solutions of the modified problem $u'' = \tilde{f}(t,u)$ are also bounded by the same R, so we can ensure that they are in fact solutions of the original problem. Let us consider some basic examples.

Example 1.1. Let $f : [0,1] \times \mathbb{R} \to \mathbb{R}$ of class be continuous and locally Lipschitz in its second variable and assume there exists a positive constant $R > 0$ such that

$$f(t,-R) < 0 < f(t,R) \qquad \text{for all } t \in [0,1]. \tag{1.4}$$

Then the Dirichlet problem (1.1)–(1.2) has at least one solution u with $\|u\|_\infty \leq R$.

Condition (1.4) is a particular case of the so-called *Hartman condition*, which shall be introduced subsequently. Here, it seems reasonable to define

$$\tilde{f}(t,u) := \begin{cases} f(t,u) & \text{if } |u| \leq R, \\ f(t,R) & \text{if } u > R, \\ f(t,-R) & \text{if } u < -R. \end{cases}$$

This "cutoff" operation might look a bit drastic; nevertheless, it is true that $\tilde{f}$ is still continuous and locally Lipschitz in u. Moreover, it is bounded, so the shooting method provides a solution u of the problem $u''(t) = \tilde{f}(t,u)$ satisfying (1.2). Thus, it suffices to prove that $|u(t)| \leq R$ for all t. Suppose, for example, that u achieves its maximum value at some t_0 with $u(t_0) > R$; then $t_0 \in (0,1)$ and

$$u''(t_0) = \tilde{f}(t_0, u(t_0)) = f(t_0, R) > 0,$$

a contradiction. The proof that $u(t) \geq -R$ for all t is analogous. Note that, in this case, we have proven that the norm of an arbitrary solution of the problem with $\tilde{f}$ is bounded by R, although the original problem might also have other solutions.

Example 1.2. Let $f : [0,1] \times \mathbb{R} \to \mathbb{R}$ be continuous and locally Lipschitz in its second variable, and assume that f is nondecreasing with respect to u, that is,

$$f(t,u) \leq f(t,v) \quad \text{for all } t \in [0,1], u,v \in \mathbb{R}, u \leq v.$$

Then the Dirichlet problem (1.1)–(1.2) has a unique solution.

This problem looks more tricky, but it can still be solved using only elementary arguments. In the first place, we should observe a remarkable novelty with respect to the preceding examples: here, the solution is unique. Thus it seems like a good idea to try to understand the role of the monotonicity condition in order to see how uniqueness follows from it.

To this end, note that if u and v are solutions of the problem, then the function $w := u - v$ satisfies

$$w''(t)w(t) = [f(t,u(t)) - f(t,v(t))][u(t) - v(t)] \geq 0.$$

Moreover, $w(0) = w(1) = 0$, so integration of the previous inequality yields

$$\int_0^1 w'(t)^2 \, dt = -\int_0^1 w''(t)w(t), \ dt \le 0.$$

This implies that $w' \equiv 0$, which in turn implies $w \equiv 0$.

At this point, we might say: "Great, we have proven uniqueness, but... how do we now deduce the *existence* of solutions?" As we shall see in subsequent chapters, this is a particular case of a rather general class of problems in which, roughly speaking, *uniqueness implies existence*. As before, we shall replace f by a cutoff function $\tilde{f}$; however, in the present case we do not have, for the moment, a value R on which to cut. Thus, we must choose it in an accurate way.

That is the key idea of the aforementioned a priori bounds: sometimes it is possible, before knowing whether or not the problem has a solution, to prove that the norm of such a solution, if it does exist, is smaller than a certain constant. Here, if u solves (1.1)–(1.2), then we may write

$$u''(t) = f(t, u(t)) - f(t, 0) + f(t, 0)$$

and, since $u(0) = 0$, we obtain

$$u''(t)u(t) = [f(t, u(t)) - f(t, 0)]u(t) + f(t, 0)u(t) \ge f(t, 0)u(t).$$

Thus, integration at both sides of the inequality yields

$$\int_0^1 u'(t)^2 \, dt \le -\int_0^1 f(t, 0)u(t) \, dt \le \|u\|_\infty \int_0^1 |f(t, 0)| \, dt. \tag{1.5}$$

Next, recall the standard notation

$$A^+ = \max\{A, 0\}, \qquad A^- = \max\{-A, 0\},$$

and write $u(t) = \int_0^t u'(s) \, ds$ to deduce that

$$-\int_0^1 [u'(s)]^- \, ds \le -\int_0^t [u'(s)]^- \, ds \le u(t) \le \int_0^t [u'(s)]^+ \, ds \le \int_0^1 [u'(s)]^+ \, ds$$

for all t. Furthermore, since $u(0) = u(1)$,

$$0 = \int_0^1 u'(s) \, ds = \int_0^1 ([u'(s)]^+ - [u'(s)]^-) \, ds,$$

so

$$\int_0^1 [u'(s)]^- \, ds = \int_0^1 [u'(s)]^+ \, ds = \frac{1}{2} \int_0^1 ([u'(s)]^+ + [u'(s)]^-) \, ds = \frac{1}{2} \int_0^1 |u'(s)| \, ds.$$

From the previous inequalities, $|u(t)| \le \frac{1}{2} \int_0^1 |u'(s)| \, ds$, and hence, by the Cauchy–Schwarz inequality, we conclude that

$$\|u\|_\infty \le \frac{1}{2}\|u'\|_{L^2}.$$

Combined with (1.5), this implies

$$\|u\|_\infty \le \frac{1}{2}\|u'\|_{L^2} \le \frac{1}{2}\int_0^1 |f(t,0)|\,dt := R.$$

If we set $\tilde{f}$ exactly as in the previous example, then the modified problem has a solution u. But now comes the most interesting part: the bound R was obtained *using only the monotonicity of f*. Thus, because $\tilde{f}$ is also nondecreasing in its second variable and $\tilde{f}(t,0) = f(t,0)$, we deduce that $\|u\|_\infty$ is bounded by the same R, and hence u is a solution of the original problem.

Example 1.3. Let $f : [0,1] \times \mathbb{R} \to \mathbb{R}$ be continuous and locally Lipschitz in its second variable, and assume that f has *linear growth* in u, that is, $|f(t,u)| \le A|u| + B$ for some constants $A, B \ge 0$. We claim that, if $A < 4$, then the the Dirichlet problem (1.1)–(1.2) has a solution. Indeed, in this case, an a priori bound for an arbitrary solution u is obtained as follows. As before, let us multiply by $u(t)$ and integrate to obtain

$$\int_0^1 u'(t)^2\,dt = -\int_0^1 f(t,u(t))u(t)\,dt \le A\int_0^1 u(t)^2\,dt + B\int_0^1 |u(t)|\,dt.$$

The computations in the previous example now yield

$$\|u\|_\infty^2 \le \frac{1}{4}\|u'\|_{L^2}^2 \le \frac{A\|u\|_\infty^2 + B\|u\|_\infty}{4},$$

and hence $\|u\|_\infty \le \frac{B}{4-A}$. Fix $R \ge \frac{B}{4-A}$, and let $\tilde{f}$ be defined as previously. Moreover, $|\tilde{f}(t,u)| \le A|u| + B$, so the solutions of the truncated problem are, in particular, solutions of the original one.

Remark 1.1. The condition $A < 4$ can be improved; indeed, using the last inequality in Appendix B, Sect. B.2.2, it is verified that

$$\|u\|_\infty \le \frac{1}{8}\|u''\|_\infty \le \frac{A\|u\|_\infty + B}{8},$$

and it suffices to assume $A < 8$. Furthermore, from the methods that shall be introduced in subsequent chapters, it is possible to prove the existence of solutions under the even weaker condition $A < \pi^2$. This latter assumption is already sharp, as may be seen from the following example. Let $f(t,u) = \sin(\pi t) - \pi^2 u$, and suppose that (1.1)–(1.2) has a solution u. Multiply by $\sin(\pi t)$ at both sides and integrate by parts to obtain

$$\int_0^1 \sin^2(\pi t)\,dt = \int_0^1 [u''(t) + \pi^2 u(t)]\sin(\pi t)\,dt = 0,$$

a contradiction.

1.2 From Scalar Equations to Systems

Besides the extensions obtained in previous examples, the application of the shooting method to the case of a bounded nonlinearity might have been a bit disappointing for those expecting something really spectacular. In some sense, the resolution was just *too* simple: notice, for example, that our conclusions have relied only on the fact that if the absolute value of the first derivative of a solution u at the initial point $t = 0$ is large, then u is monotone. This is, indeed, very simple, although it is not clear how it might be extended for a *system* of equations.

The goal of this section consists in solving, instead of a scalar equation, a system of two equations. In other words, we shall still consider Eq. (1.1), but now with $f : [0,1] \times \mathbb{R}^2 \to \mathbb{R}^2$, and look for a solution $u : [0,1] \to \mathbb{R}^2$ satisfying the Dirichlet conditions (1.2).

Of course, it is natural to ask why we should restrict ourselves to the two-dimensional case. In fact, all the results in this section can be extended to higher dimensions; however, systems of two equations already contain the main ingredients of the nonscalar case and have the additional advantage of allowing a very elementary and elegant approach. Analogous conclusions for the general case can be easily obtained from the results that shall be developed later, in Chap. 4.

1.2.1 With a Little Help from My Intuition

With the scalar case still in mind, let us now assume that $f : [0,1] \times \mathbb{R}^2 \to \mathbb{R}^2$ is continuous and locally Lipschitz in u and define, for $\lambda \in \mathbb{R}^2$, the function u_λ as the unique solution of the initial value problem (1.1)–(1.3). As before,

$$u'_\lambda(t) = \lambda + \int_0^t f(s, u(s))\, ds.$$

But this time we may perform a second integration step to obtain

$$u_\lambda(t) = \lambda + \int_0^t \int_0^\tau f(s, u_\lambda(s))\, ds\, d\tau.$$

As long as u_λ is still defined for $t = 1$, we may set again $\phi(\lambda) := u_\lambda(1)$, and hence

$$\phi(\lambda) = \lambda + S,$$

where S (for "something") satisfies $|S| \le \|f\|_\infty$.

The bad news is that, in this new context, the Bolzano theorem cannot be used since we are now dealing with a two-dimensional problem. Consequently, the existence of a zero of ϕ is not yet obvious.

Let us first try a simple argument: in the first place, it is clear that if $R > \|f\|_\infty$, then $|\phi(\lambda) - \lambda| < R$ for all λ. Thus,

$$|\lambda|^2 - 2\phi(\lambda) \cdot \lambda + |\phi(\lambda)|^2 < R^2,$$

and, in particular, when $|\lambda| = R$, we deduce

$$\phi(\lambda) \cdot \lambda > \frac{1}{2}|\phi(\lambda)|^2 \geq 0.$$

Now recall that $\phi(\lambda) \cdot \lambda = |\phi(\lambda)|.|\lambda| \cos \alpha$, where α is the angle between $\phi(\lambda)$ and λ; hence, the previous inequality just says that, when $|\lambda| = R$, the vector field ϕ points outwards from the ball $\overline{B_R(0)}$. Is this enough to conclude that ϕ vanishes in $B_R(0)$?

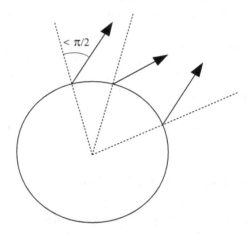

Suppose it does not; then for each fixed $\lambda \in \overline{B_R(0)}$ the line $L\delta := \lambda + \delta\phi(\lambda)$ hits the circumference $\partial B_R(0)$ at a unique nonnegative value of $\delta = \delta(\lambda)$. More precisely, from the equality

$$|\lambda + \delta\phi(\lambda)|^2 = R^2$$

we obtain

$$\delta(\lambda) = \frac{\phi(\lambda) \cdot \lambda + \sqrt{(\phi(\lambda) \cdot \lambda)^2 + R^2 - |\lambda|^2}}{|\phi(\lambda)|^2}.$$

Moreover, because $\phi(\lambda) \cdot \lambda > 0$ for $|\lambda| = R$, the term inside the square root in the previous expression is strictly positive; thus, the function δ is as smooth as ϕ.

We conclude that the mapping $r : \overline{B_R(0)} \to \mathbb{R}^2$ defined by

$$r(\lambda) = \lambda + \delta(\lambda)\phi(\lambda)$$

is smooth and has two very special properties:

1. $r(\overline{B_R(0)}) \subset \partial B_R(0)$,

2. $\lambda \in \partial B_R(0) \Rightarrow r(\lambda) = \lambda$.

In other words, r is a so-called *retraction*, that maps a closed disk onto its boundary and leaves the points of this boundary fixed. As we shall see, such a mapping cannot exist.

1.2.2 In the Beginning Was Green

In this section, we shall prove that there are no C^2 retractions from the closed unit disk onto its boundary. Later on, the result shall be extended for *continuous* retractions; in this first stage, the smoothness assumption shall be very helpful in using only elementary tools. Indeed, a remarkable aspect of the following proof is the fact that it employs just arguments from a basic course in calculus.

By contradiction, suppose $r = (u, v) : \overline{B_1(0)} \to \partial B_1(0)$ is a C^2 mapping such that $r|_{\partial B_1(0)} = id$, and define the quantity

$$d := \int_{\partial B_1(0)} (u \nabla v - v \nabla u) \cdot d\sigma.$$

On the one hand, setting $\gamma(t) := (\cos t, \sin t)$ and using the fact that $r \circ \gamma = \gamma$ we deduce, by the chain rule,

$$\nabla u \circ \gamma \cdot \gamma' = \gamma_1', \quad \nabla v \circ \gamma \cdot \gamma' = \gamma_2',$$

and hence

$$d = \int_0^{2\pi} \left[u(\gamma(t))\gamma_2'(t) - v(\gamma(t))\gamma_1'(t) \right] dt = \int_0^{2\pi} \left[\gamma_1(t)\gamma_2'(t) - \gamma_2(t)\gamma_1'(t) \right] dt$$
$$= \int_0^{2\pi} \left[\cos^2(t) + \sin^2(t) \right] dt = 2\pi.$$

On the other hand, we may write

$$d = \int_{\partial B_1(0)} P dx + Q dy,$$

where

$$P := u\frac{\partial v}{\partial x} - v\frac{\partial u}{\partial x}, \quad Q := u\frac{\partial v}{\partial y} - v\frac{\partial u}{\partial y}.$$

Because $u^2 + v^2 \equiv 1$, it is seen that

$$u\frac{\partial u}{\partial x} + v\frac{\partial v}{\partial x} = u\frac{\partial u}{\partial y} + v\frac{\partial v}{\partial y} \equiv 0.$$

In other words, ∇u and ∇v are linearly dependent, that is,

$$\frac{\partial u}{\partial x}\frac{\partial v}{\partial y} = \frac{\partial u}{\partial y}\frac{\partial v}{\partial x}.$$

Hence,

$$\frac{\partial Q}{\partial x} = \frac{\partial \left(u\frac{\partial v}{\partial y} - v\frac{\partial u}{\partial y}\right)}{\partial x} = \frac{\partial u}{\partial x}\frac{\partial v}{\partial y} - \frac{\partial u}{\partial y}\frac{\partial v}{\partial x} + u\frac{\partial^2 v}{\partial y\partial x} - v\frac{\partial^2 u}{\partial y\partial x}$$

$$= \frac{\partial u}{\partial y}\frac{\partial v}{\partial x} - \frac{\partial u}{\partial x}\frac{\partial v}{\partial y} + u\frac{\partial^2 v}{\partial x\partial y} - v\frac{\partial^2 u}{\partial x\partial y} = \frac{\partial \left(u\frac{\partial v}{\partial x} - v\frac{\partial u}{\partial x}\right)}{\partial y} = \frac{\partial P}{\partial y}$$

and by Green's theorem we conclude that $d = 0$, a contradiction.

Remark 1.2. Note that the explicit computation of $d = 2\pi$ can be performed without invoking any specific parameterization of $\partial B_1(0)$. Indeed, for arbitrary γ we may apply Green's theorem again to obtain

$$d = \int_0^{2\pi} \left[\gamma_1(t)\gamma_2'(t) - \gamma_2(t)\gamma_1'(t)\right] dt = \int_\gamma x\,dy - y\,dx = \iint_{B_1(0)} 2\,dx\,dy = 2\pi.$$

Thus we have proven the following theorem.

Theorem 1.1. *There is no C^2 mapping $r : \overline{B_1(0)} \to \partial B_1(0)$ such that $r|_{\partial B_1(0)} = \mathrm{id}$.*

As a consequence of this "smooth" version of the no-retraction theorem, we also deduce, from the computations in the previous section, that if ϕ is a C^2 mapping from the closed unit ball that points outward over the boundary, then it necessarily vanishes in the interior.

Theorem 1.2. *Assume that $\phi : \overline{B_1(0)} \to \mathbb{R}^2$ is a C^2 mapping and satisfies*

$$\phi(x) \cdot x > 0 \qquad \text{for } x \in \partial B_1(0). \tag{1.6}$$

Then ϕ has a zero in $B_1(0)$.

Furthermore, we also obtain a C^2 version of a very famous result.

Theorem 1.3. *(Brouwer's theorem, C^2 version) Let $f : \overline{B_1(0)} \to \overline{B_1(0)}$ be a C^2 mapping. Then f has at least one fixed point, that is, there exists $x \in \overline{B_1(0)}$ such that $f(x) = x$.*

Proof. If $f(x) = x$ for some $x \in \partial B_1(0)$, then we are done. Otherwise, define the C^2 mapping $\phi(x) := x - f(x)$; then, for $x \in \partial B_1(0)$,

$$\phi(x) \cdot x = |x|^2 - f(x) \cdot x = 1 - f(x) \cdot x > 0$$

since $f(x) \neq x$ and the norm of both is less than or equal to 1. Thus, ϕ vanishes in $B_1(0)$. $\square$

Add More Dimensions to Your Bolzano: The Poincaré–Miranda Theorem

A few pages ago, we complained about the fact that Bolzano's theorem could not be used for a two-dimensional shooting problem. Now we must confess that we were not being completely sincere; indeed, the following result can be regarded as a generalization of Bolzano's theorem for vector fields in $\mathbb{R}^2$.

Theorem 1.4. *(Poincaré–Miranda, C^2 version) Let $f : [-1,1] \times [-1,1] \to \mathbb{R}^2$ be a C^2 mapping such that*

$$f_1(-1,x_2) < 0 < f_1(1,x_2) \qquad \forall x_2 \in [-1,1] \qquad (1.7)$$

and

$$f_2(x_1,-1) < 0 < f_2(x_1,1) \qquad \forall x_1 \in [-1,1]. \qquad (1.8)$$

Then there exists $x \in (-1,1) \times (-1,1)$ such that $f(x) = (0,0)$.

This result, also known as *the generalized intermediate value theorem*, can be proven in several ways. For example, one may consider the function $g(x) := x - \frac{f(x)}{M}$, where M is a constant. It is readily seen that, for M large enough, g maps the square $[-1,1] \times [-1,1]$ into a smaller square $C \subset (-1,1) \times (-1,1)$. Next, take a C^2-diffeomorphism $d : \overline{B_1(0)} \to K$ for some K such that $C \subset K \subset [-1,1] \times [-1,1]$. In particular, $g(K) \subset K$, so by Theorem 1.3 the mapping $d^{-1} \circ g \circ d$ has a fixed point y. Then $x = d(y)$ is a fixed point of g and consequently, a zero of f.

Remark 1.3. It is clear that both the generalized intermediate value theorem and Theorem 1.2 are still valid if any of the inequalities (1.6), (1.7), or (1.8) is reversed. Furthermore, these inequalities do not need to be strict; indeed, assume, for example, that $\phi : \overline{B_1(0)} \to \mathbb{R}^2$ is a C^2 mapping such that $\phi(x) \cdot x \geq 0$ on $\partial B_1(0)$, and define $\phi_n(x) := \phi(x) + \frac{x}{n}$. Then $\phi_n(x) \cdot x > 0$ over $\partial B_1(0)$, so ϕ_n has a zero x_n. By compactness, there exists a subsequence of $\{x_n\}$ that converges to some $x \in \overline{B_1(0)}$, which is obviously a zero of ϕ. A similar argument allows us to prove a nonstrict version of Theorem 1.4.

1.2.3 *Green Light (to Topology)*

As shown in the previous section, there are no C^2 retractions from the closed unit ball in $\mathbb{R}^2$ onto its boundary; in particular, this property was used to prove Theorems 1.2–1.4.

This would suffice for our immediate purposes: if we are just dealing with a simple system like (1.1), it does no harm to assume that the involved functions are smooth, so the mapping $\phi(\lambda) := u_\lambda(1)$ is also smooth (Appendix B.1, Exercise 3b). However, it requires no great effort to prove that the preceding "topological" results are, indeed, topological: in other words, they still hold if we assume that the mappings are only continuous. To this end, we shall make use of the Stone–Weierstrass theorem; the reader who wishes to keep using only arguments from very basic calculus may skip this part and proceed further with the remaining sections in this chapter.

Let us start with Theorem 1.2. If a continuous function $\phi : \overline{B_1(0)} \to \mathbb{R}^2$ satisfies (1.6), then we may fix a sequence of C^2 functions $\phi_n : \overline{B_1(0)} \to \mathbb{R}^2$ such that $\phi_n \to \phi$ uniformly. Let $\varepsilon := \min_{x \in \partial B_1(0)} \phi(x) \cdot x > 0$ and n_0 such that $\|\phi_n - \phi\|_\infty < \varepsilon$ for $n \geq n_0$; then

$$\phi_n(x) \cdot x = \phi(x) \cdot x - [\phi(x) - \phi_n(x)] \cdot x \geq \varepsilon - \|\phi_n - \phi\|_\infty > 0$$

for any $x \in \partial B_1(0)$ and $n \geq n_0$. Thus, ϕ_n satisfies (1.6) for n large enough, and hence it has a zero x_n. Next, take a subsequence of $\{x_n\}$ converging to some $x \in \overline{B_1(0)}$; then it follows easily that $\phi(x) = 0$.

From this point, the extension of the no-retraction theorem is straightforward: if $r : \overline{B_1(0)} \to \partial B_1(0)$ is a continuous retraction, then it satisfies (1.6), and hence it should vanish, a contradiction. Summarizing:

Theorem 1.5. *There is no continuous mapping $r : \overline{B_1(0)} \to \partial B_1(0)$ such that $r(x) = x$ for all $x \in \partial B_1(0)$.*

Theorem 1.6. *If $\phi : \overline{B_1(0)} \to \mathbb{R}^2$ is continuous such that $\phi(x) \cdot x > 0$ for $x \in \partial B_1(0)$, then ϕ has a zero in $B_1(0)$.*

Either statement can be used to prove the standard version of Brouwer's theorem in $\mathbb{R}^2$ (alternatively, one might use the Stone–Weierstrass theorem again and deduce it directly from Theorem 1.3).

Theorem 1.7. *(Brouwer) Any continuous mapping $f : \overline{B_1(0)} \to \overline{B_1(0)}$ has at least one fixed point.*

The same is true for the generalized intermediate value theorem:

Theorem 1.8. *(Poincaré–Miranda) Let $f : [-1,1] \times [-1,1] \to \mathbb{R}^2$ be a continuous mapping satisfying (1.7) and (1.8). Then there exists $x \in (-1,1) \times (-1,1)$ such that $f(x) = (0,0)$.*

Again, the proof follows from the Stone–Weierstrass theorem combined with Theorem 1.4, or just repeating the argument of the preceding section and applying Theorem 1.7. Alternative proofs can be obtained using Theorem 1.5 or 1.6. Regarding this last one, it is worth observing that if a function f satisfies (1.7) and (1.8), then it is an outward pointing field with respect to the square $[-1,1] \times [-1,1]$.

Remark 1.4. Although the four preceding theorems were introduced in a specific order, they are all equivalent: that is, any one of them can be used to prove any of the others. Furthermore, they all imply the completeness axiom for real numbers (Exercise 1.5). This is also true if we consider only the C^2 versions of the last section: thus, we may conclude that Green is *really* "in the beginning." This might seem a bit weird since a lot of work is required just to understand the statement of Green's theorem: one needs to deal with oriented curves and line integrals, double integrals, partial derivatives, and so on. But, formally, it is true that all the properties of the real numbers are valid if the completeness axiom is replaced by the statement of Green's theorem over the unit ball.

1.3 Poincaré Mapping

In this section, we shall investigate the application of the shooting method to a periodic problem. To begin, consider the first-order system

$$\begin{cases} u'(t) = f(t, u(t)), \\ u(0) = u(1), \end{cases} \tag{1.9}$$

where $f : [0,1] \times \mathbb{R}^2 \to \mathbb{R}^2$ is continuous and locally Lipschitz in its second variable $u \in \mathbb{R}^2$.

The idea of the method is essentially the same: define u_λ as the solution of the system with initial data $\lambda \in \mathbb{R}^2$, and try to find an appropriate value of λ such that u_λ solves (1.9). Specifically, u_λ is defined as the unique solution of

$$\begin{cases} u'(t) = f(t, u(t)), \\ u(0) = \lambda, \end{cases}$$

and we look for some λ such that $u_\lambda(1) = \lambda$. In other words, we search for a fixed point of the so-called *Poincaré mapping P*, defined by $P(\lambda) = u_\lambda(1)$. Again, trajectories are not necessarily defined up to $t = 1$, so the domain of P may not be the whole plane; however, in some cases it is possible to obtain enough information to ensure the existence of a fixed point in some subset of the plane. This is the case in the following elementary example.

Proposition 1.1. *Let* $f : [0,1] \times \mathbb{R}^2 \to \mathbb{R}^2$ *be a* C^2 *mapping, and assume that*

$$f(t, u) \cdot u < 0 \qquad for \ |u| = R,$$

where R is some positive constant. Then (1.9) *has at least one solution u such that* $\|u\|_\infty \leq R$.

Proof. Take $\lambda \in \mathbb{R}^2$ such that $|\lambda| \leq R$, and assume that u_λ is defined on $[0, T]$ for some T. We claim that $|u_\lambda(t)| \leq R$ for all $t \in [0, T]$. Indeed, setting $\psi(t) := |u(t)|^2$ we obtain

$$\psi'(t) = 2u(t) \cdot u'(t) = 2u(t) \cdot f(t, u(t)).$$

In particular, if $|u(t)| = R$, then $\psi'(t) < 0$. First, assume that $u(0) \in B_R(0)$; then $u(t)$ cannot reach the boundary $\partial B_R(0)$ for any t. On the other hand, if $|u(0)| = R$, then $\psi(t)$ is initially decreasing and, again, $u(t)$ must remain inside the ball $B_R(0)$ for all $t \in (0, T)$. This implies, in the first place, that u_λ is defined on $[0, 1]$ for $|\lambda| \leq R$

and, consequently, the Poincaré mapping P is well defined and C^2 (Appendix B.1). Furthermore, $P(\overline{B_R(0)}) \subset \overline{B_R(0)}$, so the existence of a fixed point of P follows from Theorem 1.3. $\square$

1.3.1 *Smoothing for Shooting

The reader might have noticed, in the last proposition, that the smoothness assumption on f was made with the specific purpose of using the C^2 version of Brouwer's theorem. This looks quite fair since a nonstarred section should use results from nonstarred sections only, although it is quite obvious that the result is still valid if f is only continuous and locally Lipschitz in u. This is not a big deal since we know that the more general Theorem 1.7 holds.

But what happens if f is only continuous? The study of this case seems to go against the essence of the shooting method, which is based on the existence and uniqueness theorem for the initial value problem. However, an extension of Proposition 1.1 can be easily obtained by means of a procedure which is, in fact, very general. In a similar fashion as in Sect. 1.2.3, we shall approximate f by smooth functions and obtain a convergent sequence of solutions of the approximated problems. In Theorem 1.6, this was almost trivial because the closed unit ball in $\mathbb{R}^2$ was compact; in the present problem, we are dealing with a sequence of *functions*, so we shall make use of a well-known and powerful compactness result: the Arzelá–Ascoli theorem. Before jumping into details, it is worth observing that even a more general result may be obtained with almost the same effort: again, the inequality for f does not need to be strict.

Corollary 1.1. *(Proposition 1.1 revisited) Let $f : [0,1] \times \mathbb{R}^2 \to \mathbb{R}^2$ be continuous, and assume that*

$$f(t,u) \cdot u \leq 0 \qquad \text{for } |u| = R,$$

where R is some positive constant. Then (1.9) has at least one solution u such that $\|u\|_\infty \leq R$.

Proof. By the Stone–Weierstrass theorem, for each n we may set a smooth mapping $f_n : [0,1] \times \mathbb{R}^2 \to \mathbb{R}^2$ such that $|f_n(t,u) - f(t,u)| < \frac{R}{n}$ for all $t \in [0,1]$ and $u \in \overline{B_R(0)}$. Next, define $g_n(t,u) := f_n(t,u) - \frac{u}{n}$; thus, if $|u| = R$, then

$$g_n(t,u) \cdot u = [f_n(t,u) - f(t,u)] \cdot u + f(t,u) \cdot u - \frac{R^2}{n} < 0.$$

From Proposition 1.1, problem (1.9) with g_n instead of f has a solution u_n such that $\|u_n\|_\infty \leq R$. Furthermore, for arbitrary t

$$|u_n'(t)| = |g_n(t, u_n(t))| \leq \max_{t \in [0,1], |u| \leq R} |g_n(t,u)| \leq K$$

for some constant K independent of n. From the Arzelá–Ascoli theorem there exists a subsequence $\{u_{n_j}\}$ that converges uniformly on $\overline{B_R(0)}$ to some function u, so if we write

$$u_{n_j}(t) = u_{n_j}(0) + \int_0^t g_{n_j}(s, u_{n_j}(s))\, ds,$$

then taking limit we deduce

$$u(t) = u(0) + \int_0^t f(s, u(s))\, ds$$

for all t, and hence u is a solution of (1.9). □

1.3.2 When Poincaré Met Miranda

Let us take again the second-order problem (1.1), with $f : [0,1] \times \mathbb{R} \to \mathbb{R}$ a C^2 function, but now under the periodic conditions

$$u(0) = u(1), \quad u'(0) = u'(1). \tag{1.10}$$

Note that the problem is scalar, although, unlike the case of Dirichlet conditions, it cannot be reduced to a one-dimensional fixed point problem. Indeed, here the Poincaré mapping has two parameters, which correspond to the initial values of the function and its derivative.

As a first example, assume the Hartman condition (1.4). Observe that the cutoff function of Sect. 1.1 does not preserve the smoothness; however, this will not be a big problem: it is easy to find a bounded C^2 function $\tilde{f}$ such that $\tilde{f} \equiv f$ over $[0,1] \times [-R,R]$ and $\tilde{f}(t,-u) < 0 < \tilde{f}(t,u)$ for all $u \geq R$ and $t \in [0,1]$. As in previous examples, if u satisfies (1.10) and solves the modified problem $u''(t) = \tilde{f}(t, u(t))$, then $\|u\|_\infty \leq R$.

Indeed, suppose that u achieves its global maximum at some t_0 with $u(t_0) > R$. As before, if $t_0 \in (0,1)$, then $u''(t_0) = \tilde{f}(t_0, u(t_0)) > 0$, a contradiction; otherwise, it follows from (1.10) that $u(0) = u(1) = \max_{0 \leq t \leq 1} u(t)$. Hence, $u'(0) \leq 0 \leq u'(1)$, and using (1.10) again we deduce that $u'(0) = 0 = u'(1)$. Moreover, $u''(t)$ is positive near $t = 0$, so u' is initially increasing. As $u'(0) = 0$, it follows that u is also initially increasing, which contradicts the fact that a global maximum is achieved at $t = 0$. Similarly, it is proven that $u(t) \geq -R$ for all t.

Thus, it suffices to verify that the modified problem has a solution. To this end, for fixed $x, y \in \mathbb{R}$ let $u_{x,y}$ be the unique solution of the initial value problem

$$u''(t) = \tilde{f}(t, u(t)),$$

$$u(0) = x, \quad u'(0) = y.$$

For convenience, instead of looking for a fixed point of the Poincaré mapping $P(x,y) := u_{x,y}(1)$, we shall try to find a zero of the function

$$F(x,y) := (u'_{x,y}(1) - y, u_{x,y}(1) - x). \tag{1.11}$$

First, note that

$$u'_{x,y}(t) = y + \int_0^t \tilde{f}(s, u_{x,y}(s)) \, ds$$

and

$$u_{x,y}(t) = x + ty + \int_0^t (t - s)\tilde{f}(s, u_{x,y}(s)) \, ds.$$

Fixing $M > \|\tilde{f}\|_\infty$, it is seen that

$$u_{x,M}(1) - x = M + \int_0^1 (1 - s)\tilde{f}(s, u_{x,y}(s)) \, ds \geq M - \|\tilde{f}\|_\infty > 0$$

for all x and, in the same way, $u_{x,-M}(1) - x < 0$.

Finally, fix $\tilde{R} := R + 2M$, and let $y \in [-M, M]$; then

$$u_{\tilde{R},y}(t) = \tilde{R} + yt + \int_0^t (t - s)\tilde{f}(s, u_{\tilde{R},y}(s)) \, ds > \tilde{R} - 2M = R.$$

Hence, we conclude from (1.4) that $\tilde{f}(t, u_{\tilde{R},y}(t)) > 0$ for all t. Thus,

$$u'_{\tilde{R},y}(1) - y = \int_0^1 \tilde{f}(t, u_{\tilde{R},y}(t)), \, dt > 0,$$

and, analogously, $u'_{-\tilde{R},y}(1) - y < 0$. The conclusion follows now from the Poincaré–Miranda theorem applied on the rectangle $[-\tilde{R}, \tilde{R}] \times [-M, M]$.

Remark 1.5. In particular, for any continuous function $a > 0$ and g bounded the problem

$$u''(t) - a(t)u(t) = g(t, u(t))$$

has a solution satisfying (1.10). Indeed, it suffices to apply the previous example with $f(t,u) := a(t)u + g(t,u)$ and $R > \frac{\|g\|_\infty}{\min_{t \in [0,1]} a(t)}$.

An interesting result in the same direction was proven by Lazer in [68]. For simplicity, let $g \in C^2(\mathbb{R}, \mathbb{R})$ be bounded and assume there exist $R > 0$ and $c \in \mathbb{R}$ such that

$$g(u) < c < g(-u) \qquad \text{for all } u \text{ such that } |u| > R \tag{1.12}$$

or

$$g(u) > c > g(-u) \qquad \text{for all } u \text{ such that } |u| > R. \tag{1.13}$$

Then the periodic problem

$$u''(t) + g(u(t)) = p(t), \qquad u(0) = u(1), \quad u'(0) = u'(1), \tag{1.14}$$

has a solution for each C^2 function p satisfying $\int_0^1 p(t) \, dt = c$.

Note that, in the present situation, the Hartman condition would read

$$g(R) < p(t) < g(-R) \qquad \text{for all } t,$$

so it is somehow comparable to (1.12). However, this is a *pointwise* inequality: the remarkable aspect of Lazer's result is the fact that the condition is given only in terms of the *average* of the function p.

The original proof used a well-known result called *Schauder's theorem*, which will be introduced in Chap. 3; the methods described in this chapter allow a much more elementary proof. With this aim, we may try to reproduce the previous argument with $\tilde{f}(t, u) := p(t) - g(u)$. It seems this is our lucky day since everything is working in exactly the same way until the inequality $u_{\tilde{R},y}(t) > R$ is deduced for all t and all $y \in [-M, M]$.

But now... wait! We cannot conclude, as before, that $\tilde{f}(t, u_{\tilde{R},y}(t)) > 0$ for all t, although if, for example, (1.12) holds, then we obtain

$$u'_{\tilde{R},y}(1) - y = \int_0^1 p(t) - g(u_{\tilde{R},y}(t)) \, dt$$

$$= c - \int_0^1 g(u_{\tilde{R},y}(t)) \, dt = \int_0^1 [c - g(u_{\tilde{R},y}(t))] \, dt > 0.$$

Similarly, it is seen that $u'_{-\tilde{R},y}(1) - y < 0$, so the Poincaré–Miranda theorem can be applied to the same mapping F defined in (1.11). The proof is similar under condition (1.12).

Remark 1.6. * The general result by Lazer is in fact established for g continuous and satisfying one of the nonstrict conditions

$$g(u) \geq c \geq g(-u) \qquad \text{for all } u \text{ such that } |u| > R \qquad (1.15)$$

or

$$g(u) \leq c \leq g(-u) \qquad \text{for all } u \text{ such that } |u| > R. \qquad (1.16)$$

Furthermore, the boundedness condition on g may also be relaxed: it suffices to assume that g is *sublinear*:

$$\lim_{|u| \to \infty} \frac{g(u)}{u} = 0. \qquad (1.17)$$

Also, a friction term can be added to the equation, namely,

$$u''(t) + au'(t) + g(u(t)) = p(t), \qquad (1.18)$$

for some constant a. In this case, the integral expression for $u(t)$ and $u'(t)$ is a bit different, but it can be easily obtained by the method of variation of parameters. An interesting fact is that, when $a \neq 0$, the sublinearity assumption on g can be dropped; the same is true when (1.16) holds instead of (1.15).

This general result can be deduced from the preceding case using the ideas of Sects. 1.1.2 and 1.3.1; to illustrate this fact, let us obtain a priori bounds when $a \neq 0$

under condition (1.12) or (1.13) and leave the rest of the proof and the remaining cases as an exercise for the reader. Assume that u is a solution of (1.18) such that $u(0) = u(1)$ and $u'(0) = u'(1)$. Multiplying the equation by u' and integrating, we obtain

$$\int_0^1 [u''(t)u'(t) + au'(t)^2 + g(u(t))u'(t)] \, dt = \int_0^1 p(t)u'(t) \, dt.$$

Set $G(u) := \int_0^u g(s) \, ds$; then

$$\int_0^1 [u''(t)u'(t) + g(u(t))u'(t)] \, dt = \left[\frac{u'^2}{2} + G(u) \right]\Bigg|_0^1 = 0,$$

and thus

$$\int_0^1 u'(t)^2 \, dt = \frac{1}{a} \int_0^1 p(t)u'(t) \, dt \leq \frac{1}{|a|} \|p\|_{L^2} \|u'\|_{L^2}.$$

Hence,

$$\|u'\|_{L^2} \leq \frac{\|p\|_{L^2}}{|a|} := r.$$

On the other hand, if we write $u(t) - u(0) = \int_0^t u'(s) \, ds$, then it follows, as in Example 1.2, that

$$|u(t) - u(0)| \leq \frac{1}{2} \|u'\|_{L^2} \leq \frac{r}{2}$$

for all t. Finally, integrate (1.18) and use the periodic conditions (1.10) to obtain

$$\int_0^1 g(u(t)) \, dt = \int_0^1 p(t) \, dt = c.$$

In particular, there exists a value $t_0 \in (0,1)$ such that $g(u(t_0)) = c$. We claim that $|u(0)| \leq R + \frac{r}{2}$ since otherwise $|u(t)| \geq |u(0)| - |u(t) - u(0)| > R$ for all t, and hence $g(u(t)) \neq c$ for all t, a contradiction.

Thus, $|u(t)| \leq |u(t) - u(0)| + |u(0)| \leq R + r$ for all t.

1.4 *Let's Make It Complex: The Index of a Curve

Although all the Bolzano-like arguments of this chapter can be extended for higher dimensions, we have considered only the two-dimensional case because it enables a very simple and intuitive treatment using only Green's theorem. In this section, we shall introduce an even simpler argument that requires just a little bit more: basic complex analysis. In fact, none of that is needed for the C^2 case, as the following elementary proof of the fixed point theorem shows.

Let $f : \overline{B_1(0)} \to \overline{B_1(0)}$ be a C^2 mapping, and fix a C^2 function $\lambda : [0,2] \to [0,1]$ such that $\lambda \equiv 1$ on $[0, \frac{1}{2}]$ and $\lambda \equiv 0$ on $[1,2]$. Next, identify $\mathbb{R}^2$ with $\mathbb{C}$, and define $h : [0, 2\pi] \times [0,1] \to \mathbb{C}$ by

$$h(t,s) = \lambda(s)e^{it} - (1 - \lambda(2s))f(\lambda(s)e^{it}).$$

Suppose that f has no fixed points; then h does not vanish. Indeed,

- If $s \geq \frac{1}{2}$, then $h(t,s) = \lambda(s)e^{it} - f(\lambda(s)e^{it}) \neq 0$ since f has no fixed points;
- If $s < \frac{1}{2}$, then $h(t,s) = e^{it} - (1 - \lambda(2s))f(e^{it}) \neq 0$ since $\|f\|_\infty \leq 1$ and, again, f has no fixed points.

Thus, we may define

$$I(s) := \int_0^{2\pi} \frac{\frac{\partial h}{\partial t}(t,s)}{h(t,s)} \, dt. \tag{1.19}$$

By standard results I is differentiable, and

$$I'(s) = \int_0^{2\pi} \frac{\partial}{\partial s}\left(\frac{\frac{\partial h}{\partial t}(t,s)}{h(t,s)}\right) dt = \int_0^{2\pi} \frac{\partial}{\partial t}\left(\frac{\frac{\partial h}{\partial s}(t,s)}{h(t,s)}\right) dt = \left.\frac{\frac{\partial h}{\partial s}(\cdot,s)}{h(\cdot,s)}\right|_0^{2\pi}.$$

Now observe that $h(0,s) = h(2\pi,s)$ for all s; in other words, the curve $\gamma(t) := \frac{\frac{\partial h}{\partial s}(t,s)}{h(t,s)}$ is closed, and hence $I'(s) = 0$ for all s. On the other hand,

- $h(t,0) = e^{it}$, so $I(0) = 2\pi i$;
- $h(t,1) \equiv -f(0)$, so $I(1) = 0$.

This contradicts the fact that I is constant.

As mentioned, the previous proof required no complex analysis at all. Complex notation was used for the sake of convenience; the only fact that a distrustful reader should need to check is that the differentiation of curves satisfies the quotient rule, which is hidden in the middle step:

$$\left(\frac{\gamma(t)}{\varphi(t)}\right)' = \frac{\gamma'(t)\varphi(t) - \gamma(t)\varphi'(t)}{\varphi(t)^2}.$$

Also, there was perhaps a "dubious" step that involved differentiation under an integral sign, but this is easily justified in this case since the integrand is a C^1 function.

Nevertheless, the reader who is familiar with the techniques of complex analysis has surely identified the key idea behind the preceding proof, which is based on two concepts:

1. The index or *winding number* of a continuous closed curve $\gamma : [a,b] \to \mathbb{C}\backslash\{0\}$, defined by

$$I(\gamma,0) := \frac{1}{2\pi i} \int_\gamma \frac{1}{z} \, dz \in \mathbb{Z},$$

In particular, if γ is of class C^1, then $I(\gamma,0) = \frac{1}{2\pi i} \int_a^b \frac{\gamma'(t)}{\gamma(t)} \, dt$. The meaning of the previous integral when γ is just continuous might look a bit mysterious, but it is easily defined since the integrand is analytic using, for example, a so-called *primitive along the curve.*

2. *Homotopy invariance*: if $\gamma, \delta : [a,b] \to \mathbb{C}\backslash\{0\}$ are continuous closed curves and $h : [a,b]\times \to \mathbb{C}\backslash\{0\}$ is continuous and satisfies $h(t,0) = \gamma(t)$, $h(t,1) = \delta(t)$ for all t and $h(a,s) = h(b,s)$ for all s, then

$$I(\gamma,0) = I(\delta,0).$$

Remark 1.7. The reader who hates paragraphs starting with "the reader who..." may employ the previous ideas for a better understanding of the last two concepts. On the one hand, it is possible to obtain an alternative proof of the homotopy invariance of the index when the homotopy of class C^2: simply observe that the integral $I(s)$ in (1.19) is $2\pi i$ times the index of the curve $h(\cdot,s)$ around 0, and the conclusion $I'(s) = 0$ is valid for arbitrary h, so any pair of homotopic curves necessarily have the same winding number.

On the other hand, even the definition of index and the fact that it is an integer can be explained, when the curve is smooth, without invoking the theory of complex analysis. For instance, let $\gamma : [0,1] \to \mathbb{C}\backslash\{0\}$ be a closed C^1 curve, and assume for simplicity that its image intersects the y-axis only finitely many times. Next, write $\gamma(t) = x(t) + iy(t)$, and compute

$$\frac{\gamma'(t)}{\gamma(t)} = \frac{\gamma'(t)\overline{\gamma(t)}}{|\gamma(t)|^2} = \frac{x'(t)x(t) + y'(t)y(t)}{x(t)^2 + y(t)^2} + \frac{y'(t)x(t) - x'(t)y(t)}{x(t)^2 + y(t)^2}i.$$

The real part of this expression is the derivative of the function $\frac{1}{2}\ln(x(t)^2 + y(t)^2)$, which has the same value at $t = 0$ and $t = 1$, so the imaginary part of $I(\gamma,0)$ vanishes. Moreover, the imaginary part of $\frac{\gamma'}{\gamma}$ also has an easy-to-compute primitive, namely, $\arctan\frac{y(t)}{x(t)}$. But, unlike the previous case, this function is not defined when $x(t) = 0$, so we need to split the integral as the sum of several integrals over smaller intervals. From our assumptions, the set $\{t \in [0,1] : x(t) = 0\}$ is finite; moreover, observe that if $t_0 < t_1$ are two consecutive zeros of x, then

$$\int_{t_0}^{t_1} \frac{y'(t)x(t) - x'(t)y(t)}{x(t)^2 + y(t)^2}\, dt = \arctan\frac{y(t)}{x(t)}\bigg|_{t_0}^{t_1} = \lim_{t \to t_1^-}\arctan\frac{y(t)}{x(t)} - \lim_{t \to t_0^+}\arctan\frac{y(t)}{x(t)}.$$

As the sign of $x(t)$ between t_0 and t_1 is constant, we have three possible cases:

1. If $sgn(y(t_0)) = sgn(y(t_1))$, then $\int_{t_0}^{t_1} \frac{y'(t)x(t) - x'(t)y(t)}{x(t)^2 + y(t)^2}\, dt = 0$.
2. If, for example, $y(t_0) < 0 < y(t_1)$ and $x(t) > 0$ on (t_0, t_1), then

$$\int_{t_0}^{t_1} \frac{y'(t)x(t) - x'(t)y(t)}{x(t)^2 + y(t)^2}\, dt = \lim_{u \to +\infty}\arctan u - \lim_{u \to -\infty}\arctan u = \pi.$$

The same result is obtained if $y(t_0) > 0 > y(t_1)$ and $x(t) < 0$ on (t_0, t_1).
3. An analogous calculation shows that the result is $-\pi$ in the two remaining situations:

   - $y(t_0) < 0 < y(t_1)$ and $x(t) < 0$ on (t_0, t_1).
   - $y(t_0) > 0 > y(t_1)$ and $x(t) > 0$ on (t_0, t_1).

Thus, every arc between two consecutive zeros of x contributes to the integral $I(\gamma,0) = \frac{1}{2\pi i} \int \frac{\gamma'(t)}{\gamma(t)} dt$ with 0, $\frac{1}{2}$, or $-\frac{1}{2}$ according to the three preceding cases; because γ is closed, it is clear that the total number of arcs where case 2 or 3 occurs is even, so $I(\gamma,0)$ is an integer.

Such a machinery allows a more comprehensive proof of Brouwer's theorem in two simple steps, for example, proving first the following lemma.

Lemma 1.1. *Let* $f : \overline{B_1(0)} \to \mathbb{C}\setminus\{0\}$ *be continuous, and let* $\gamma(t) = e^{it}$ *for* $t \in [0, 2\pi]$. *Then* $I(f \circ \gamma, 0) = 0$.

Proof. It suffices to consider the homotopy $h(t,s) = f(s\gamma(t))$. $\square$

Using this lemma, the proof of the fixed point theorem (and some of its generalizations) is straightforward: suppose that $f : \overline{B_1(0)} \to \overline{B_1(0)}$ is continuous with $f(z) \neq z$ for all z, and define $g(z) = z - f(z)$. On the one hand, consider the homotopy $h(t,s) = \gamma(t) - sf(\gamma(t))$. Suppose $h(t,s) = 0$; then $|sf(\gamma(t))| = |\gamma(t)| = 1$, so $s = 1$ and $f(\gamma(t)) = \gamma(t)$, a contradiction. Hence, $I(g \circ \gamma, 0) = I(\gamma, 0) = 1$. On the other hand, the previous lemma says that $I(g \circ \gamma, 0) = 0$, so a contradiction arises.

1.4.1 Rock Around the Clock... and Find Many Solutions

Besides a great variety of topological applications, the winding number can also be used in the theory of differential equations, for example, to prove the existence of solutions to a rather general class of boundary value problems. For simplicity, let us consider a particular example:

$$u''(t) + u(t)^3 = p(t), \tag{1.20}$$

with $p : [0,1] \to \mathbb{R}$ continuous, under the Dirichlet conditions (1.2). As was done earlier, we define $u_\lambda(t)$ as the unique solution of the initial value problem (1.20)–(1.3). Multiplication by $u'_\lambda(t)$ and integration yield

$$u'_\lambda(t)^2 + \frac{u_\lambda(t)^4}{2} = \lambda^2 + 2\int_0^t p(s)u'_\lambda(s)\, ds, \tag{1.21}$$

so for $|\lambda|$ large we deduce that

$$u'_\lambda(t)^2 \leq r\lambda^2 \text{ and } u_\lambda(t)^4 \leq r\lambda^2 \tag{1.22}$$

for some constant $r > 1$. For example, if $|\lambda| \geq \int_0^1 |p(t)|\, dt$, then by direct computation we obtain $u'_\lambda(t)^2 \leq (1+\sqrt{2})^2\lambda^2$ and, consequently, $u_\lambda(t)^4 \leq [2+4(1+\sqrt{2})]\lambda^2$.

In particular, it follows that u_λ is defined on $[0,1]$. Moreover, if t_0 is a critical point of u, then, enlarging $|\lambda|$ as much as necessary,

$$u_\lambda(t_0)^4 \geq 2\lambda^2 - 4\int_0^1 |p(t)u_\lambda'(t)|\, dt > \|p\|_\infty^{4/3}.$$

Hence, from (1.20)

$$u''(t_0) = p(t_0) - u(t_0)^3 < 0 \qquad \text{if } u(t_0) > 0$$

and

$$u''(t_0) = p(t_0) - u(t_0)^3 > 0 \qquad \text{if } u(t_0) < 0.$$

In other words, when λ is large, all local maxima are positive and local minima are negative. From Rolle's theorem, we conclude that zeros and critical points of u alternate: and there are no other critical points. More precisely, if we define $\mathscr{C}_\lambda :=$ $\{t \in [0,1] : u_\lambda(t) = 0 \text{ or } u_\lambda'(t) = 0\}$, then

$$\mathscr{C}_\lambda = \{0 = t_0 < t_1 < \ldots < t_N\}$$

for some $N = N(\lambda)$ and, for $j = 1, \ldots, N$,

$$u_\lambda(t_j) = 0 \neq u_\lambda'(t_j) \ \text{ if } j \text{ is even}$$

$$u_\lambda(t_j) \neq 0 = u_\lambda'(t_j) \ \text{ if } j \text{ is odd}.$$

Furthermore, when $j \leq N - 2$ is odd, $u_\lambda(t_j)$ and $u_\lambda(t_{j+2})$ have opposite signs.

Following Remark 1.7, let us consider now the curve $\gamma_\lambda : [0,1] \to \mathbb{C} \setminus \{0\}$ given by $\gamma_\lambda(t) := u_\lambda'(t) + iu_\lambda(t)$ and the integral

$$I(\lambda) := \frac{1}{2\pi} \int_0^1 \frac{u_\lambda'(t)^2 - u_\lambda''(t)u_\lambda(t)}{u_\lambda'(t)^2 + u_\lambda(t)^2}\, dt = \frac{1}{2\pi} \int_0^1 \frac{u_\lambda'(t)^2 + u_\lambda(t)^4 - p(t)u_\lambda(t)}{u_\lambda'(t)^2 + u_\lambda(t)^2}\, dt.$$

When $|\lambda|$ is large, the integrand is positive, so γ_λ rotates counterclockwise and $I(\lambda)$ measures the net fraction of turns that the curve performs around the origin. In consequence,

$$I(\lambda) = \tfrac{n}{2} \text{ with } n \in \mathbb{N} \iff N = 2n \text{ and } t_N = 1$$

$$\iff u_\lambda \text{ is a solution of (1.20)–(1.2) with } \textit{exactly } n - 1 \text{ zeros in } (0,1).$$

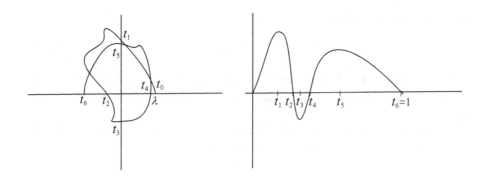

We claim that the distance between any pair of consecutive points of $\mathscr{C}_\lambda$ tends to 0 as $|\lambda| \to \infty$; in particular, this implies that $N(\lambda) \to \infty$ and, hence, $I(\lambda) \to +\infty$ as $|\lambda| \to \infty$. Moreover, I is continuous, so we deduce the existence of some $y_0 \geq 0$ such that I takes all the values $y \geq y_0$ at least two times. Hence, for any $n \in \mathbb{N}$ with $n \geq 2y_0$ there exist at least two solutions of (1.20)–(1.2) having exactly $n-1$ zeros in $(0,1)$.

To prove the claim, let us verify, for example, that $t_1 - t_0 \to 0$ as $\lambda \to +\infty$; the other cases are similar. In $(0,t_1)$, both $u(t)$ and $u'(t)$ are positive, and from (1.21) it follows that $u'_\lambda(t)^2 \geq \frac{\lambda^2 - u_\lambda(t)^4}{2}$ when λ is sufficiently large. Thus, if we define the value $t_\lambda := \sup\{t \in [0,t_1] : u_\lambda(t)^4 \leq \lambda^2\}$, then we obtain

$$\frac{u'_\lambda(t)}{\sqrt{\lambda^2 - u_\lambda(t)^4}} \geq \frac{\sqrt{2}}{2} \text{ for } t < t_\lambda,$$

and consequently

$$\int_0^{t_\lambda} \frac{u'_\lambda(t)}{\sqrt{\lambda^2 - u_\lambda(t)^4}} dt \geq \frac{\sqrt{2}}{2} t_\lambda.$$

Because

$$\int_0^{t_\lambda} \frac{u'_\lambda(t)}{\sqrt{\lambda^2 - u_\lambda(t)^4}} dt = \int_0^{u_\lambda(t_\lambda)} \frac{du}{\sqrt{\lambda^2 - u^4}} \leq \frac{1}{\sqrt{\lambda}} \int_0^1 \frac{dv}{\sqrt{1 - v^4}},$$

it follows that $t_\lambda \leq \frac{C}{\sqrt{\lambda}}$ for some constant C. Finally, for $t \in [t_\lambda, t_1]$ it holds that $u_\lambda(t) \geq \sqrt{\lambda}$, and thus

$$u'_\lambda(t) = -\int_t^{t_1} u''_\lambda(s)\, ds = \int_t^{t_1} [u_\lambda(s)^3 - p(s)]\, ds \geq \lambda^{3/2}(t_1 - t) - \int_0^1 |p(t)|\, dt,$$

which in turn implies

$$u_\lambda(t_1) - u_\lambda(t_\lambda) = \int_{t_\lambda}^{t_1} u'_\lambda(s)\, ds \geq \lambda^{3/2} \frac{(t_1 - t_\lambda)^2}{2} - (t_1 - t_\lambda) \int_0^1 |p(t)|\, dt.$$

From the second inequality of (1.22) it is seen that

$$(r^{1/4} - 1)\lambda^{1/2} \geq \lambda^{3/2} \frac{(t_1 - t_\lambda)^2}{2} - (t_1 - t_\lambda) \int_0^1 |p(t)|\, dt.$$

Finally, enlarge λ if necessary to conclude that $t_1 - t_\lambda \leq \frac{K}{\sqrt{\lambda}}$ for some constant K, which completes the proof.

Appendix

A few historical notes and further comments. The shooting method was introduced in 1905 by Severini [106], although, as mentioned in [47], the modern version of the method has its origins in a more sophisticated technique, known as the *Ważewski method*, which makes use of a topological lemma closely related to Brouwer's theorem (e.g., [44]). Historical aspects of the two-point boundary value problems for ordinary equations are presented in [80].

As mentioned, the case $n = 2$ of Brouwer's theorem is very special since its proof, far from being trivial as in the case $n = 1$, can still be performed using only basic tools. For instance, it is very easy to conclude that there are no retractions from $B_1(0)$ to S^1 since otherwise there would be an epimorphism from the fundamental group of $B_1(0)$—which is trivial—onto $\mathbb{Z}$, the fundamental group of the unit circumference. When $n > 2$, the fundamental group of S^{n-1} is also trivial, so more information is needed. Elementary arguments are also possible, as seen, in the context of complex analysis and—amazingly—in game theory: indeed, the result can be deduced from the so-called Hex theorem, which, roughly speaking, establishes that any game of Hex has a winner [39]. It is also worth mentioning that Brouwer's theorem allows one to prove another of the best known results in the topology of the plane, which can also be generalized for higher dimensions: the Jordan curve theorem.

Problems

One-Dimensional Shooting

1.1. Prove that the forced pendulum equation with friction

$$u''(t) + au'(t) + b\sin(u(t)) = p(t) \tag{1.23}$$

with $p : [0, 1] \to \mathbb{R}$ continuous admits at least one solution for the arbitrary Dirichlet conditions

$$u(0) = u_0, \qquad u(1) = u_1.$$

Does an analogous result hold for the periodic conditions

$$u(0) = u(1), \qquad u'(0) = u'(1)?$$

Generalize for the problem $u''(t) + au'(t) = f(t, u(t))$ with f bounded.

1.2. Let $g : \mathbb{R} \to \mathbb{R}$ be a C^1 and bounded function such that

$$g(0) = 0 \qquad \text{and} \qquad [(2k-1)\pi]^2 < g'(0) < [2k\pi]^2$$

for some integer k. Prove that the Dirichlet problem

$$u''(t) + g(u(t)) = 0, \qquad u(0) = u(1) = 0$$

has at least two different nontrivial solutions.

Hint: consider as in Sect. 1.1 the differentiable function $\phi : \mathbb{R} \to \mathbb{R}$ given by $\phi(\lambda) = u_\lambda(1)$. Then $\phi(0) = 0$ and $\phi(-\lambda) < 0 < \phi(\lambda)$ for $\lambda > \|g\|_\infty$. Next, prove that $\phi'(0) = w_0(1)$, where w_0 is the unique solution of the linear problem

$$\begin{cases} w_0''(t) + g'(0)w_0(t) = 0, \\ w_0(0) = 0, \quad w_0'(0) = 1, \end{cases}$$

and verify that $\phi'(0) < 0$. Now draw a graph of ϕ and deduce that it has at least two nontrivial zeros.

1.3. *(Strict upper and lower solutions)* Let $f : [0,1] \times \mathbb{R} \to \mathbb{R}$ be continuous and locally Lipschitz on u, and assume there exist functions $\alpha, \beta : [0,1] \to \mathbb{R}$ such that

$$\alpha(t) \le \beta(t), \qquad \alpha''(t) > f(t, \alpha(t)), \qquad \beta''(t) < f(t, \beta(t))$$

for all t and

$$\alpha(0), \alpha(1) \le 0 \le \beta(0), \beta(1).$$

Prove that the Dirichlet problem (1.1)–(1.2) admits at least one solution u with $\alpha(t) \le u(t) \le \beta(t)$ for all t. In particular, set $\alpha \equiv -R$ and $\beta \equiv R$ to obtain the Hartman condition (1.4).

Brouwer's Theorem and Related Results

1.4. Prove that all the theorems of Sect. 1.2.3 are equivalent, that is, that any of them can be used to prove any of the others. Moreover, prove that all of them are equivalent to the following statement: if $f : \mathbb{R}^2 \to \mathbb{R}^2$ is continuous and there exists a constant C such that $|f(x) - x| \le C$ for all x, then f has at least one zero.

1.5. Prove that the completeness axiom of the real numbers can be replaced by any of the statements mentioned in problem 1.4. Prove that, furthermore, it suffices to assume that all the mappings involved are of class C^2.

1.6. Use the lemma in Sect. 1.4 to give a direct proof of each of the statements mentioned in problem 1.4.

1.7. *(Hartman condition)* Let $f : [0,1] \times \mathbb{R}^2 \to \mathbb{R}^2$ be a C^1 function, and assume there exists a positive constant R such that

$$f(t, u) \cdot u > 0 \qquad \text{for all } t \in [0,1], u \in \partial B_R(0).$$

Prove that the Dirichlet problem

$$\begin{cases} u''(t) = f(t,u(t)) \\ u(0) = u_0, \quad u(1) = u_1 \end{cases} \tag{1.24}$$

has at least one solution for any $u_0, u_1 \in \overline{B_R(0)}$.

1.8. (*Monotonicity condition*) Let $f : [0,1] \times \mathbb{R}^2 \to \mathbb{R}^2$ be a C^1 function, and assume that

$$[f(t,u) - f(t,v)] \cdot (u - v) > 0$$

for all $(t,u),(t,v) \in [0,1] \times \mathbb{R}^2$. Prove that problem (1.24) has a unique solution for any $u_0, u_1 \in \mathbb{R}^2$.

1.9. *Extend problems 1.7 and 1.8 for f continuous and nonstrict inequalities.

Poincaré Mapping

1.10. Let $f : \mathbb{R} \times \mathbb{R}^2 \to \mathbb{R}^2$ be a smooth function, T-periodic in t and sublinear in x, that is,

$$f(t + T, x) = f(t,x) \qquad \text{for all } (t,x),$$

$$\lim_{|x| \to \infty} \frac{f(t,x)}{|x|} = 0 \qquad \text{uniformly on } t.$$

1. Prove that the problem

$$x'(t) + x(t) = f(t,x(t))$$

admits at least one T-periodic solution.
2. Prove that the problem

$$x'(t) - x(t) = f(t,x(t))$$

admits at least one T-periodic solution.

1.11. Let $f,g : \mathbb{R} \times \mathbb{R}^2 \to \mathbb{R}$ be of class C^1 and T-periodic in t. Furthermore, assume there exist $R_1, R_2 > 0$ such that

$$f(t,R_1,y)f(t,-R_1,y) < 0$$

for each $t \in \mathbb{R}$ and $y \in \mathbb{R}$ such that $|y| \leq R_2$, and

$$g(t,x,R_2)g(t,x,-R_2) < 0$$

for each $t \in \mathbb{R}$ and $x \in \mathbb{R}$ such that $|x| \leq R_1$. Prove that the problem

$$\begin{cases} x' = f(t,x(t),y(t)) \\ y' = g(t,x(t),y(t)) \end{cases}$$

admits at least one T-periodic solution.

1.12. *(Adapted from [46]). Prove the following *Lazer–Leach theorem* [70]: let $g :$ $\mathbb{R} \to \mathbb{R}$ be continuous and bounded, and assume that the limits

$$g(\pm\infty) := \lim_{u \to \pm\infty} g(u)$$

exist. Furthermore, let $p \in C(\mathbb{R}, \mathbb{R})$ be 2π-periodic, and define

$$A := \int_0^{2\pi} p(t) \cos(t) \, dt, \qquad B := \int_0^{2\pi} p(t) \sin(t) \, dt.$$

Then the problem

$$u''(t) + u(t) + g(u(t)) = p(t)$$

has at least one 2π-periodic solution, provided that

$$\sqrt{A^2 + B^2} < 2|g(+\infty) - g(-\infty)|. \tag{1.25}$$

Hint: first, assume that g is smooth, and define as in Sect. 1.3.2 the mapping

$$F(x,y) := (y - u'_{x,y}(2\pi), u_{x,y}(2\pi) - x).$$

Next, use the polar coordinates $(x, y) := (r\cos(\theta), r\sin(\theta))$ in order to prove that if r is large enough, then $F(x,y) \cdot (x,y) \neq 0$ for $(x,y) \in \partial B_r(0)$, so the result follows. The result for g continuous is deduced by an approximation argument.

Chapter 2
The Banach Fixed Point Theorem

2.1 Brief Introduction to Fixed Point Methods

As shown in the first chapter, the shooting method makes it possible to reduce some boundary value problems to a matter of finding the zeros or fixed points of a certain function of one or more real variables. But, more generally, a wide range of problems in nonlinear analysis may be presented in the form of an abstract equation

$$Lu = Nu,$$

where $L : X \to Z$ is a linear operator and $N : Y \to Z$ is a nonlinear operator defined in appropriate normed spaces $X \subset Y$ and Z. For example, the Dirichlet problem (1.1)–(1.2) fits into this setting, with $Lu := u''$ and $Nu := f(\cdot, u)$. In this case, N is defined, for example, over the set of continuous functions, but L requires twice differentiable functions. Together with the boundary conditions, this yields the following possible choice of X, Y and Z:

$$X = \{u \in C^2([0,1]) : u(0) = u(1) = 0\},$$

$$Y = \{u \in C([0,1]) : u(0) = u(1) = 0\},$$

$$Z = C([0,1]).$$

In some cases, one may just consider the restriction of N to X and try to find zeros of the function $F : X \to Z$ given by $Fu = Lu - Nu$; however, in many situations this approach is not enough, and a different analysis is required. In particular, the previous Dirichlet problem is an example of so-called *nonresonant problems* since the operator $L : X \to Z$ is invertible: for each $\varphi \in Y$, the problem $u''(t) = \varphi(t)$ has a unique solution $u \in X$. Thus, the functional equation $Lu = Nu$ is transformed into a fixed point problem:

$$u = L^{-1}Nu.$$

P. Amster, *Topological Methods in the Study of Boundary Value Problems*, Universitext, DOI 10.1007/978-1-4614-8893-4_2, © Springer Science+Business Media New York 2014

Several abstract tools have been developed to deal with problems of this kind; in this chapter, we begin with one of the most popular fixed point theorems in complete metric spaces—the contraction mapping theorem.

2.2 A Well-Known Case: Picard Method of Successive Approximations

In the first chapter, we made extensive use of the well-known existence and uniqueness theorem for ordinary differential equations.

Theorem 2.1. *Let Ω be an open subset of $\mathbb{R} \times \mathbb{R}^n$, and let $(t_0, x_0) \in \Omega$ and $f : \Omega \to \mathbb{R}^n$ be continuous and locally Lipschitz in x. Then there exists $\delta > 0$ such that the initial value problem*

$$\begin{cases} x'(t) = f(t, x(t)) \\ x(t_0) = x_0 \end{cases} \tag{2.1}$$

has a unique solution defined on $[t_0 - \delta, t_0 + \delta]$.

The standard proof of this result is based on the so-called *Picard method* of successive approximations, which consists in defining the sequence

$$x_0(t) \equiv x_0,$$

$$x_{n+1}(t) = x_0 + \int_{t_0}^{t} f(s, x_n(s)) ds$$

and showing, for some appropriate $\delta > 0$, that x_n is well defined and converges uniformly on $[t_0 - \delta, t_0 + \delta]$ to some function x. Taking the limit in the previous inequality yields

$$x(t) = x_0 + \int_{t_0}^{t} f(s, x(s)) ds,$$

and hence x is a solution of (2.1). Uniqueness follows easily from *Gronwall's lemma* (see Exercise 2 in Appendix B.1). This lemma yields, in fact, a stronger (nonlocal) uniqueness result: if x and y are solutions of (2.1) defined over an arbitrary open interval I containing t_0, then $x(t) = y(t)$ for all $t \in I$.

In his doctoral thesis from 1920, Stefan Banach showed [5] that the Picard method is, in fact, a particular case of a much more general result: the *contraction mapping principle*, also known as the *Banach fixed point theorem*.

2.3 Contraction Mapping Theorem and Its Applications

Let X and Y be metric spaces. A mapping $T : X \to Y$ is called a *contraction* if it is globally Lipschitz with constant $\alpha < 1$. In other words, $T : X \to Y$ is a contraction if there exists $\alpha < 1$ such that $d(Tx_1, Tx_2) \leq \alpha d(x_1, x_2)$ for all $x_1, x_2 \in X$. Note that we used the same d for the distance in both X and Y; this is not a problem, in particular, because we shall only consider the case $X = Y$.

Theorem 2.2. *(Banach) Let X be a complete metric space, and let $T : X \to X$ be a contraction. Then T has a unique fixed point $\hat{x}$. Furthermore, if x_0 is an arbitrary point of X and a sequence is defined iteratively by $x_{n+1} = T(x_n)$, then $\hat{x} = \lim_{n \to \infty} x_n$.*

The classical picture of this result is a map, placed precisely on the region it represents: unless the scale is 1:1 (which may sound great in J. L. Borges stories, but it is not very common in cartography), then there is exactly one point that coincides with its representation on the map.

The standard proof of the Banach theorem recalls somewhat Zeno's argument of Achilles and the tortoise: the Greek hero starts at x_0 and the tortoise at x_1, then each moves according to the rule given by T. The distance between them gets smaller as n becomes large; more precisely, $d(x_{n+1}, x_n) \leq \alpha^n d(x_1, x_0)$. Hence, for all k

$$d(x_{n+k}, x_n) \leq \sum_{j=n}^{n+k-1} d(x_{j+1}, x_j) \leq \sum_{j=n}^{n+k-1} \alpha^j d(x_1, x_0) = \frac{\alpha^n - \alpha^{n+k}}{1 - \alpha} d(x_1, x_0),$$

which implies that $\{x_n\}$ is a Cauchy sequence. The completeness of X ensures that there is no hole through which the tortoise can escape: the preceding sequence, no matter where we choose to start, converges to a point $\hat{x} \in X$. As $T(x_n) = x_{n+1} \to \hat{x}$, the continuity of T implies that $T(\hat{x}) = \hat{x}$; moreover, if also $T(y) = y$ for some $y \in X$, then $d(y, \hat{x}) = d(T(y), T(\hat{x})) \leq \alpha d(y, \hat{x})$, and we deduce that $d(y, \hat{x}) = 0$, that is: $y = \hat{x}$.

The previous computations also make it possible to estimate the rate of convergence:

$$d(x_n, \hat{x}) = \lim_{k \to \infty} d(x_n, x_{n+k}) \leq \frac{\alpha^n}{1 - \alpha} d(x_1, x_0).$$

There exists, however, an even more elementary proof of the theorem, due to R. Palais [99], that does not even require knowing how to sum up a geometric progression. First, observe that, for any $x, y \in X$,

$$d(x, y) \leq d(x, T(x)) + d(T(x), T(y)) + d(y, T(y)),$$

and hence

$$(1 - \alpha)d(x, y) \leq d(x, T(x)) + d(y, T(y)).$$

Now take $x = x_n$, $y = x_{n+k}$:

$$d(x_n, x_{n+k}) \leq \frac{1}{1 - \alpha} \left[d(T^n(x_0), T^n(x_1)) + d(T^{n+k}(x_0), T^{n+k}(x_1)) \right]$$

$$\leq \frac{\alpha^n + \alpha^{n+k}}{1 - \alpha} d(x_0, x_1).$$

Although less sharp than the previous one, this bound shows, again, that $\{x_n\}$ is a Cauchy sequence and so converges to some $\hat{x}$. Furthermore, the same rate of convergence is obtained:

$$d(x_n,\hat{x}) = \lim_{k \to \infty} d(x_n, x_{n+k}) \leq \lim_{k \to \infty} \frac{\alpha^n + \alpha^{n+k}}{1 - \alpha} d(x_0, x_1) = \frac{\alpha^n}{1 - \alpha} d(x_1, x_0).$$

As happens in another famous history of Achilles and the tortoise [24], someone may not be completely convinced yet about the validity of our proofs, so we provide yet another proof, simply based on the quantity

$$A := \inf_{x \in X} d(x, T(x)).$$

Suppose $A > 0$, and fix an arbitrary $x \in X$ such that $d(x, T(x)) < 2A$. Take n such that $\alpha^n < 2$, and deduce, by induction,

$$d(T^n(x), T(T^n(x))) = d(T^n(x), T^{n+1}(x)) \leq \alpha^n d(x, T(x)) < A.$$

This contradicts the fact that A is the infimum; thus, $A = 0$. Now take a sequence $\{x_n\}$ such that $d(x_n, T(x_n)) \to 0$. As before, it is seen that

$$d(x_n, x_{n+k}) \leq \frac{d(x_n, T(x_n)) + d(x_{n+k}, T(x_{n+k}))}{1 - \alpha},$$

which tends to 0 as $n \to \infty$. Again, we deduce that $\{x_n\}$ converges to some $\hat{x}$, and clearly $d(\hat{x}, T(\hat{x})) = 0$.

The contraction mapping theorem allows a simple and direct proof of the Picard existence and uniqueness theorem. In this case, we want to solve the functional equation

$$x(t) = x_0 + \int_0^t f(s, x(s)) \, ds,$$

so the "obvious" fixed point operator is

$$Tx(t) := x_0 + \int_0^t f(s, x(s)) \, ds.$$

We only need to find an appropriate complete metric space X such that $T : X \to X$ is well defined and contractive.

To this end, let us first consider constants $\hat{\delta}, r > 0$, such that $K \subset \Omega$, where

$$K := [t_0 - \hat{\delta}, t_0 + \hat{\delta}] \times \overline{B_r(x_0)}.$$

Next, define $M := \|f|_K\|_\infty$ and L as the Lipschitz constant of f over K, and let

$$X := \{x \in C([t_0 - \delta, t_0 + \delta], \mathbb{R}^n) : x(t) \in \overline{B_r(x_0)} \text{ for all } t\}$$

for some $\delta \leq \hat{\delta}$ to be established. In other words, X is just the closed ball of radius r centered in x_0 in the space $C([t_0 - \delta, t_0 + \delta], \mathbb{R}^n)$, equipped with the usual metric

$$d(x, y) = \max_{t \in [t_0 - \delta, t_0 + \delta]} |x(t) - y(t)|.$$

It is clear that $T : X \to C([t_0 - \delta, t_0 + \delta], \mathbb{R}^n)$ is well defined and, moreover,

$$|Tx(t) - x_0| = \left| \int_{t_0}^{t} f(s, x(s)) ds \right| \leq M\delta.$$

Choosing $\delta \leq \frac{r}{M}$, it follows that X is an invariant set, i.e., $T(X) \subset X$. On the other hand for $x, y \in X$, then

$$d(Tx, Ty) = \max_{t \in [t_0 - \delta, t_0 + \delta]} \left| \int_{t_0}^{t} (f(s, x(s)) - f(s, y(s))) ds \right| \leq \delta L d(x, y).$$

Hence, it suffices to take $\delta < \min\{\frac{r}{M}, \frac{1}{L}\}$, and then $T : X \to X$ is a contraction.

Although the Banach theorem ensures that the fixed point is unique, an extra step is needed for the uniqueness invoked in Theorem 2.1 since, in principle, for the same δ there might be other solutions that abandon the ball $\overline{B_r(x_0)}$. One possible line of reasoning is as follows: suppose y is another solution and fix $\overline{\delta} \in (0, \delta]$ such that $|y(t)| \leq r$ for $t \in [t_0 - \overline{\delta}, t_0 + \overline{\delta}]$. Next, redefine the space X accordingly, so the operator T has a unique fixed point and thus $x = y$ on $[t_0 - \overline{\delta}, t_0 + \overline{\delta}]$. This proves only *local* uniqueness, in the sense that two solutions must coincide in a neighborhood of t_0. But now the same existence and (local) uniqueness result does the rest of the job: suppose two solutions x and y are defined over an open interval I containing t_0; then the set $J := \{t \in I : x(t) = y(t)\}$ is closed in I and nonempty. Moreover, if $t_1 \in J$, then $x(t_1) = y(t_1)$, and hence $x \equiv y$ in a neighborhood of t_1. This shows that J is open and, consequently, $J = I$.

A simple corollary of the Banach theorem is as follows.

Theorem 2.3. *Let X be a complete metric space, and let $T : X \to X$ be a mapping. If $T^n := T \circ T \circ \cdots \circ T$ (n times) is a contraction for some positive integer n, then T has a unique fixed point.*

Proof. Let x be the unique fixed point of T^n; then $T^n(T(x)) = T(T^n(x)) = T(x)$. Thus, $T(x)$ is a fixed point of T^n, and, by uniqueness, we deduce that $T(x) = x$. Moreover, if y is also a fixed point of T, then $T^n(y) = y$, and, using uniqueness again, it follows that $y = x$. $\square$

A typical application of this extension is a well-known result used somewhat in the preceding chapter, if $f : [a, b] \times \mathbb{R}^n \to \mathbb{R}^n$ is continuous and globally Lipschitz with respect to x, then the solutions of all initial value problems are defined over the whole interval $[a, b]$. More precisely:

Proposition 2.1. *Let $f : [a, b] \times \mathbb{R}^n \to \mathbb{R}^n$ be continuous and globally Lipschitz with respect to x with constant L. Then for any $(t_0, x_0) \in (a, b) \times \mathbb{R}^n$ the unique solution of the problem*

$$\begin{cases} x'(t) = f(t, x(t)) \\ x(t_0) = x_0 \end{cases}$$

is defined over $[a, b]$.

An elementary proof can be obtained as an application of Gronwall's lemma (see Exercise 2 in Appendix B). In fact, such a function satisfies $|f(t,x)| \leq L|x| + c$ for some constant c, and thus solutions are globally defined. But a direct proof is also possible if we employ, as we did earlier, the operator $T : C([a,b], \mathbb{R}^n) \to C([a,b], \mathbb{R}^n)$ given by

$$Tx(t) = x_0 + \int_{t_0}^{t} f(s, x(s)) ds.$$

Indeed,

$$|T^{n+1}x(t) - T^{n+1}y(t)| = \left| \int_{t_0}^{t} [f(s, T^n x(s)) - f(s, T^n y(s))] ds \right|$$

$$\leq L \int_{a}^{t} |T^n x - T^n y|(s) ds$$

and, in particular, for $n = 0$

$$|Tx(t) - Ty(t)| \leq L \int_{a}^{t} |x(s) - y(s)| \, ds \leq L(t-a) \|x - y\|_\infty.$$

Inductively, it follows that

$$|T^n x(t) - T^n y(t)| \leq \frac{L^n (t-a)^n}{n!} \|x - y\|_\infty,$$

and hence $\|T^n x - T^n y\|_\infty \leq \frac{L^n (b-a)^n}{n!} \|x - y\|_\infty$ for all n. Now recall (for example, use the d'Alembert criterion) that $\frac{C^n}{n!} \to 0$ for any fixed constant $C > 0$, so the latter inequality implies that T^n is contractive for n large enough.   □

Example 2.1. The Banach theorem can also be applied to a second-order equation with Dirichlet conditions. The idea is simple: for fixed $v \in C([0,1])$, consider the linear problem

$$\begin{cases} u''(t) = f(t, v(t)), \\ u(0) = u(1) = 0. \end{cases} \tag{2.2}$$

If f is continuous, then (2.2) has a unique solution $u \in C^2([0,1])$, and hence the operator $T : C[0,1] \to C[0,1]$ given by

$$Tv = u, \text{ where } u \text{ is the unique solution of (2.2)},$$

is well defined. Furthermore, given $v_1, v_2 \in C[0,1]$ and $u_i = Tv_i$, it is seen that

$$(u_1 - u_2)''(t) = f(t, v_1(t)) - f(t, v_2(t)),$$
$$(u_1 - u_2)(0) = (u_1 - u_2)(1) = 0.$$

Also, we know from the appendix (Sect. B.2.2) that

$$\|u_1 - u_2\|_\infty \leq \frac{1}{8} \|u_1'' - u_2''\|_\infty;$$

in consequence, $\|u_1 - u_2\|_\infty \le \frac{1}{8}\|f(\cdot,v_1) - f(\cdot,v_2)\|_\infty$. At this point, we may assume, for example, that f is Lipschitz in the variable u with constant L and then, for all t,

$$|f(t,v_1(t)) - f(t,v_2(t))| \le L|v_1(t) - v_2(t)| \le L\|v_1 - v_2\|_\infty.$$

In particular, if $L < 8$, then T is a contraction and the problem has a unique solution. A similar result was obtained by Picard [100] in 1893, by the method of successive approximations. The proof is also valid for a *system* of second-order equations, and many other extensions are possible: for example, the left-hand side of the equation may be replaced by a more general operator or the boundary conditions may be changed (Exercise 2.2).

Remark 2.1. 1. The previous proof is readily adapted to the nonhomogeneous condition

$$u(0) = u_0, \qquad u(1) = u_1.$$

In this case, the operator $\varphi \mapsto u$, (where $u'' = \varphi$ and $u(0) = u_0$, $u(1) = u_1$) is no longer linear; however, it is affine, and the problem can be still included in the class of abstract equations $Lu = Nu$ mentioned in the introduction. It suffices to define $\theta(t) = (u_1 - u_0)t + u_0$ and $v(t) = u(t) - \theta(t)$, and thus the new problem for v reads

$$v''(t) = g(t,v(t)), \qquad v(0) = v(1) = 0,$$

where $g(t,v) := f(t,v + \theta(t))$. Note that g satisfies the same Lipschitz condition as f.

2. Sharper bounds may be obtained if we use the space $L^2(0,1)$ instead of $C([0,1])$. Indeed, if $v_1, v_2 \in L^2(0,1)$ and $u_i = Tv_i$ for $i = 1,2$, then it follows from the *Poincaré inequality* (Appendix B, Sect. B.2.2) that

$$\|u_1 - u_2\|_{L^2} \le \frac{1}{\pi}\|u_1' - u_2'\|_{L^2} \le \frac{1}{\pi^2}\|u_1'' - u_2''\|_{L^2} \le \frac{L}{\pi^2}\|v_1 - v_2\|_{L^2}.$$

Hence, it suffices to assume that $L < \pi^2$. This certainly constitutes an improvement to the previous result, although there is a possible objection to the procedure: at first sight, it is not clear how Tv is defined when v is just an element of $L^2(0,1)$. However, everything can be done in an appropriate way since—because f is Lipschitz continuous in u—it is easily seen that $f(\cdot,v) \in L^2(0,1)$, and hence it is possible to integrate it twice in the Lebesgue sense to obtain some sort of "solution" u to problem (2.2). In fact, only the first integral requires Lebesgue theory: the integral of an element of L^2 is already a continuous function, so after a second integration step we obtain a C^1 function u. Note that u shall not be in general a C^2 function, but from well-known results it is seen that u' is *absolutely continuous* so it does indeed have a derivative almost everywhere. This class of functions defines a Sobolev space called $H^2(0,1)$, which is very useful in applications and, as in the present case, allows improvements of many results in subsequent chapters.

The preceding example shows that the contraction mapping theorem is an efficient tool for proving existence and uniqueness, although its application might also be quite restrictive. The assumption that f is globally Lipschitz is already strong; furthermore, we have required the Lipschitz constant to be small. Nevertheless, there are many cases in which this assumption can be relaxed. In Picard's fundamental theorem, this was easy: only a local Lipschitz assumption was required since we were looking for local solutions; in other situations, the global Lipschitz condition may be avoided if one is able to obtain a priori bounds for the solutions, as we saw in the first chapter. But, still, one must prove that the fixed point operator is contractive: this explains why the Lipschitz constant must be small. The next result, due to Zarantonello (e.g., [23]) shows that the smallness assumption can be dropped when the operator has some other "nice" properties, such as monotonicity.

Theorem 2.4. *Let H be a Hilbert space, and assume that $T : H \to H$ is globally Lipschitz and monotone nonincreasing, that is,*

$$\langle Tx - Ty, x - y \rangle \leq 0$$

for all $x, y \in H$. Then for each fixed $y \in H$ the equation $x = Tx + y$ has a unique solution. In particular, T has a unique fixed point.

Proof. Uniqueness is clear from the monotonicity of T: indeed, if x_1 and x_2 are solutions of the equation, then $x_1 - x_2 = Tx_1 - Tx_2$, and hence

$$\|x_1 - x_2\|^2 = \langle x_1 - x_2, Tx_1 - Tx_2 \rangle \leq 0,$$

so $x_1 = x_2$. To prove existence, define, for $\lambda \in (0, 1]$, the operator

$$T_\lambda x := (1 - \lambda)x + \lambda(Tx + y).$$

Because $\lambda \neq 0$, it is clear that x is a fixed point of T_λ if and only if $x = Tx + y$. Moreover, if $x_1, x_2 \in H$, then

$$\|T_\lambda x_1 - T_\lambda x_2\|^2 = \langle T_\lambda x_1 - T_\lambda x_2, T_\lambda x_1 - T_\lambda x_2 \rangle$$

$$= (1 - \lambda)^2 \|x_1 - x_2\|^2 + \lambda^2 \|Tx_1 - Tx_2\|^2 + 2\lambda(1 - \lambda)\langle Tx_1 - Tx_2, x_1 - x_2 \rangle$$

$$\leq (1 - \lambda)^2 \|x_1 - x_2\|^2 + \lambda^2 \|Tx_1 - Tx_2\|^2 \leq [(1 - \lambda)^2 + \lambda^2 L^2] \|x_1 - x_2\|^2,$$

where L is the Lipschitz constant of T. Next, observe that the factor $(1 - \lambda)^2 + \lambda^2 L^2$, as a function of λ, achieves its absolute minimum at $\hat{\lambda} = \frac{1}{1 + L^2}$, with

$$(1 - \hat{\lambda})^2 + \hat{\lambda}^2 L^2 = \frac{L^2}{1 + L^2} < 1.$$

Thus, $T_{\hat{\lambda}}$ is a contraction. $\square$

Example 2.2. * As the reader may have noticed, the monotonicity assumption in Theorem 2.4 seems to fit perfectly well in a situation studied in the previous chapter: the Dirichlet problem $u''(t) = f(t, u(t))$ with f continuous, locally Lipschitz and nondecreasing in the second variable. In the first place, recall that (via a priori bounds *plus* truncation) it may be assumed that f is in fact globally Lipschitz. In the second place, we need a Hilbert space and a fixed point operator. The choice of the latter seems quite obvious since, as before, for any fixed function v we may solve the linear problem (2.2) and define $Tv := u$. But the choice of the appropriate Hilbert space is not so evident yet. For example, $L^2(0, 1)$ does not work: although (as mentioned in Remark 2.1) the definition of T is still correct, the monotonicity of T fails since there is no way to prove that

$$\int_0^1 (Tv_1(t) - Tv_2(t))(v_1(t) - v_2(t)) \, dt \le 0$$

for all $v_1, v_2 \in L^2(0, 1)$. So we better calm down and say to ourselves the kind of thing we learned from our grandparents: there is always something to learn from failure. In this case, the difficulty arises in connection with the fact that we do not know anything about Tv_i except that its second derivative is equal to $f(\cdot, v_i)$. And the previous integral contained no derivatives. It is different if we start instead from the equality

$$\int_0^1 (Tv_1 - Tv_2)''(t)(v_1(t) - v_2(t)) \, dt = \int_0^1 [f(t, v_1(t)) - f(t, v_2(t))](v_1(t) - v_2(t)) \, dt$$

since here we may use the very good information we have: because f is nondecreasing, the last term is nonnegative. A further step, always effective, is integration by parts:

$$\int_0^1 (Tv_1 - Tv_2)''(t)(v_1(t) - v_2(t)) \, dt = -\int_0^1 (Tv_1 - Tv_2)'(t)(v_1 - v_2)'(t) \, dt.$$

So we have the clue we needed to define the inner product of our space: if

$$\langle u, v \rangle := \int_0^1 u'(t)v'(t) \, dt,$$

then it follows from the previous computations that

$$\langle Tv_1 - Tv_2, v_1 - v_2 \rangle \le 0$$

for all functions v_1 and v_2. Now comes a new difficulty. Until now, everything was clear: the inequality holds for all functions v_1 and v_2, but where do these function live? Or, more precisely: taking into account the previous "inner product," which set of functions should we take in order to define a Hilbert space? In the first place, observe that u and v must be, in some sense, differentiable, so let us consider $C^1([0, 1])$. Is $\langle \cdot, \cdot \rangle$ an inner product for this space? The answer to this question is (obviously!) negative; in fact, all the properties are verified except *nondegeneracy*: in this context, $\langle u, u \rangle = 0$ does not imply $u = 0$. However, this can be easily fixed since the equality $\int_0^1 u'(t)^2 \, dt = 0$ certainly implies that u is a constant. Because we are looking for solutions that satisfy the homogeneous Dirichlet condition, there is no harm in assuming that the elements of the space vanish at the boundary:

$$X := \{u \in C^1([0,1]) : u(0) = u(1) = 0\}.$$

Now we have a space equipped with an inner product, so we can ask: what about completeness? At this stage, there is some bad news and some good news. The bad news is that (as anyone might have expected) X is not complete. The good news is that we don't care. This might sound a bit rude, but what we mean in fact is that we may work in a space H defined as the (unique up to isomorphisms) completion of X, without worrying about what its elements look like.

Indeed, for $v_1, v_2 \in X$ the operator T satisfies

$$\langle Tv_1 - Tv_2, Tv_1 - Tv_2 \rangle = \int_0^1 (Tv_1 - Tv_2)'(t)^2 \, dt$$

$$= -\int_0^1 (Tv_1 - Tv_2)''(t)(Tv_1 - Tv_2)(t) \, dt \le \|(Tv_1 - Tv_2)''\|_{L^2} \|Tv_1 - Tv_2\|_{L^2},$$

so using the definition of T, the Lipschitz condition, and the Poincaré inequality, we deduce that

$$\int_0^1 (Tv_1 - Tv_2)'(t)^2 \, dt \le L \|v_1 - v_2\|_{L^2} \|Tv_1 - Tv_2\|_{L^2}$$

$$\le \frac{L}{\pi} \|v_1 - v_2\|_{L^2} \|(Tv_1 - Tv_2)'\|_{L^2}$$

and, finally, that

$$\|Tv_1 - Tv_2\| = \|(Tv_1 - Tv_2)'\|_{L^2} \le \frac{L}{\pi} \|v_1 - v_2\|_{L^2}.$$

In particular, $T : X \to X$ is uniformly continuous, and hence it can be extended to a uniform continuous function (still denoted T) from H to H in a unique way.

The remaining part is quite predictable: one easily verifies that $T : H \to H$ is Lipschitz and nondecreasing, so it has a unique fixed point u. Next, we use density: take $u_n \in X$ such that $u_n \to u$; then $Tu_n \to Tu = u$. By definition, it follows that $(Tu_n)''(t) = f(t, u_n(t))$, and from here it is not difficult to verify (exercise!) that u is in fact an element of X and solves the problem.

Note, incidentally, that if $v \in X$, then $\|v\|_\infty \le \frac{1}{2} \|v'\|_{L^2} = \frac{1}{2} \|v\|$; in particular, Cauchy sequences in $(X, \|\cdot\|)$ converge uniformly. In other words, we do in fact have an idea of what the elements of H look like: for instance, they can always be identified with a continuous function. The reader who is familiar with Sobolev spaces must have obviously noticed that the "mysterious" space H is simply $H_0^1(0,1)$; we did not mention this fact explicitly in order to keep the exposition at a very elementary level. In Exercise 2.8, an alternative approach to the previous problem without using Sobolev spaces at all is introduced.

2.4 Implicit Function Theorem

In this section, we shall give a proof of one of the greatest results in mathematics: the implicit function theorem. The main ideas of this important theorem can be traced back to Newton and Leibniz, although its first rigorous statement and proof are due to Cauchy. The standard version we learn in our first courses of calculus, involving several independent and dependent real variables, was established later in 1878 by Dini (for historical and many other interesting aspects of the theorem, see [57]).

Let us start by an informal presentation of the simplest case, for a smooth function $f : U \to \mathbb{R}$, where U is an open subset of $\mathbb{R}^2$. If $f(t_0, x_0) = 0$ and $\frac{\partial f}{\partial x}(t_0, x_0) \neq 0$, then there exist a neighborhood I of t_0 and a (unique) smooth function $\varphi : I \to \mathbb{R}$ such that $\varphi(t_0) = x_0$ and $f(t, \varphi(t)) = 0$ for all $t \in I$. For a rapid proof, just observe that such a function, assuming it exists, must verify the equality

$$\frac{\partial f}{\partial t}(t, \varphi(t)) + \frac{\partial f}{\partial x}(t, \varphi(t))\varphi'(t) = 0.$$

Now take $W \subset U$ open such that $(t_0, x_0) \in W$ and $\frac{\partial f}{\partial x}(t, x) \neq 0$ for $(t, x) \in W$, and let $g : W \to \mathbb{R}$ be defined by

$$g(t, x) := -\frac{\frac{\partial f}{\partial t}(t, x)}{\frac{\partial f}{\partial x}(t, x)}.$$

From the existence and uniqueness theorem, there exists a function φ defined in an open interval containing t_0 such that $\varphi'(t) = g(t, \varphi(t))$ and $\varphi(t_0) = x_0$. It follows from the definition that the derivative of the function $\xi(t) := f(t, \varphi(t))$ is 0 for all t, so ξ is constant. Because $\xi(t_0) = f(t_0, x_0) = 0$, the proof is complete.

It may be argued that the preceding proof fails if f is just a C^1 function since the existence and uniqueness theorem requires g to be Lipschitz continuous in x; however, as shown in Exercise 2.5, the existence part of Picard's theorem is still valid without the Lipschitz assumption (see also Chap. 4). This intuitive approach can be extended for the more general n-dimensional case (see [57] for details), and in fact, essentially the same argument can be applied to prove a more general case, for C^k mappings in Banach spaces. Amazingly, this version of the implicit function theorem can be used to prove a rather general existence-uniqueness result for differential equations, so in some sense both results are different sides of the same coin. This is not exactly true, but, as we shall see, in the proofs of both results there is a common ingredient: the contraction mapping theorem.

The procedure is simple, although it requires the more specific concept of *uniform contraction*:

Definition 2.1. Let X and Y be normed spaces, and let $U \subset X$, $V \subset Y$ be open. A continuous mapping $f : U \times \overline{V} \to \overline{V}$ is called a uniform contraction if there exists $\alpha < 1$ such that

$$\|f(x, y_1) - f(x, y_2)\| \leq \alpha \|y_1 - y_2\|$$

for all $x \in U$, $y_1, y_2 \in \overline{V}$.

In particular, if $\overline{V}$ is complete, then for each $x \in U$ there exists a unique $y = y(x)$ such that $f(x,y) = y$. It is immediately verified that a function $y : U \to \overline{V}$ defined in this way is continuous: indeed, writing $y(x) = y$, $y(\hat{x}) = \hat{y}$ we obtain

$$\|y - \hat{y}\| = \|f(x,y) - f(\hat{x},\hat{y})\| \leq \|f(x,y) - f(x,\hat{y})\| + \|f(x,\hat{y}) - f(\hat{x},\hat{y})\|$$

$$\leq \alpha\|y - \hat{y}\| + \|f(x,\hat{y}) - f(\hat{x},\hat{y})\|.$$

Thus,

$$\|y - \hat{y}\| \leq \frac{1}{1 - \alpha}\|f(x,\hat{y}) - f(\hat{x},\hat{y})\| \to 0$$

as $x \to \hat{x}$.

It is worth mentioning that the full version of the implicit function theorem requires in fact a stronger result, which states that if f is a C^k mapping, then y is also a C^k mapping. However, in our applications we shall be concerned only with the *existence* of the implicit function and not with its smoothness, so we leave this result for the appendix. In any case, the formulation of the theorem and our proof of the existence part requires a little bit of calculus in normed spaces, so we shall give a brief introduction to the basic notions in the following pages.

Unless we are in serious danger of getting confused, the norm of all spaces shall be denoted by $\|\cdot\|$; in particular, we recall that if X and Y are normed spaces, then the set $L(X,Y)$ of continuous linear transformations from X to Y is also a normed space, with the standard norm defined by $\|T\| := \sup_{\|x\| \leq 1} \|Tx\|$.

Definition 2.2. Let X, Y be normed spaces, and let U be an open subset of X. A function $f : U \to Y$ is called (Fréchet) differentiable at $x_0 \in U$ if there exists a continuous linear transformation $T : X \to Y$ such that

$$R(x) := f(x) - f(x_0) - T(x - x_0)$$

satisfies $\frac{R(x)}{\|x-x_0\|} \to 0$ as $x \to x_0$.

The operator $T \in L(X,Y)$ is called the differential of f at x_0 and denoted, as usual, by $Df(x_0)$. A function f is said to be differentiable if it is differentiable at every point of U; in this case, a mapping $Df : U \to L(X,Y)$ is induced. If Df is continuous, then it is said that f is a C^1 function. The following well-known result is left as a simple exercise.

Lemma 2.1. *(Chain rule) Let X, Y, and Z be normed spaces and $U \subset X$, $V \subset Y$ be open. If $f : U \to V$ and $g : V \to Z$ are C^1 mappings, then $g \circ f : U \to Z$ is also a C^1 mapping, and $D(g \circ f)(x) = Dg(f(x)) \circ Df(x)$.*

It proves convenient, in the context of the implicit function theorem, to introduce the concept of partial derivatives. Let X, Y, Z be normed spaces, and let $f : U \times V \to Z$ be a differentiable function, where $U \subset X$ and $V \subset Y$ are open. For $(x,y) \in U \times V$, define $\frac{\partial f}{\partial x}(x,y) \in L(X,Z)$ and $\frac{\partial f}{\partial y}(x,y) \in L(Y,Z)$ by

$$\frac{\partial f}{\partial x}(x,y)(h) := Df(x,y)(h,0) = \lim_{t\to 0}\frac{f(x+th,y)-f(x,y)}{t},$$

$$\frac{\partial f}{\partial y}(x,y)(k) := Df(x,y)(0,k) = \lim_{t\to 0}\frac{f(x,y+tk)-f(x,y)}{t}.$$

Also, the following generalization of Lagrange's theorem shall be very useful.

Lemma 2.2. *(Mean value theorem) Let $f : U \subset X \to Y$ be a C^1 mapping, and assume that the segment*

$$[x_1,x_2] := \{tx_2 + (1-t)x_1 : 0 \le t \le 1\}$$

is contained in U. Then

$$\|f(x_2)-f(x_1)\| \le M\|x_1-x_2\|,$$

where $M := \max_{x\in[x_1,x_2]}\|Df(x)\|$.

Despite its simplicity, this lemma is far from being trivial, even in the finite-dimensional case: for instance, one might be tempted to use the standard Lagrange theorem in each coordinate, but this does not provide the best possible value for the constant M. However, an easy proof reads as follows: given an arbitrary element $\varphi \in L(Y,\mathbb{R})$ (the dual space of Y), we may define the function $\xi : [0,1] \to \mathbb{R}$ by $\xi(t) = \varphi \circ f(tx_2 + (1-t)x_1)$. Then

$$\varphi[f(x_2)-f(x_1)] = \xi(1) - \xi(0) = \xi'(\hat{t})$$

for some mean value $\hat{t} \in (0,1)$. Moreover, from the chain rule and the linearity of φ we deduce that

$$\xi'(\hat{t}) = \varphi[Df(\hat{t}x_2 + (1-\hat{t})x_1)(x_2-x_1)],$$

and thus

$$|\varphi[f(x_2)-f(x_1)]| \le M\|\varphi\|\|x_2-x_1\|.$$

Now we apply the Hahn–Banach theorem. Specifically, we shall need only one of its immediate consequences, which ensures, in particular, that the dual space of a nontrivial space always contains nontrivial elements.

Lemma 2.3. *Let Y be a normed space and $y \in Y\setminus\{0\}$. Then there exists $\varphi \in L(Y,\mathbb{R})$ such that $\|\varphi\| = 1$ and $\varphi(y) = \|y\|$.*

From this lemma (which can be proven in a direct way using Zorn's lemma) the proof of the mean value theorem follows by taking $y = f(x_2) - f(x_1)$ in the previous inequality. An alternative proof for those readers who prefer to avoid the use of the Hahn–Banach theorem is proposed in Exercise 2.7.

The mean value theorem has an obvious consequence: if $\|Df(x)\| \le \alpha < 1$ for all x, then f is a contraction. The following converse is always true: if a differentiable mapping f is also a contraction with constant α, then $\|Df(x)\| \le \alpha$ for all x. Indeed, it suffices to compute, for arbitrary v, the directional derivative

$$Df(x)(v) = \lim_{t \to 0} \frac{f(x+tv) - f(x)}{t},$$

and the conclusion $\|Df(x)(v)\| \le \alpha \|v\|$ follows.

Now we are able to prove the existence part of the implicit function theorem.

Theorem 2.5. *Let X, Y, and Z be Banach spaces, let $U \subset X$, $V \subset Y$ be open, and let $f : U \times V \to Z$ be a C^k mapping. Furthermore, assume that $f(x_0, y_0) = 0$ and that $\frac{\partial f}{\partial y}(x_0, y_0) : Y \to Z$ is an isomorphism. Then there exists a U_0 neighborhood of x_0 and a unique C^k function $\phi : U_0 \to V$ such that $\phi(x_0) = y_0$ and $f(x, \phi(x)) = 0$ for all $x \in U_0$.*

Proof. Define the function $g : U \times V \to Z$ by

$$g(x, y) := y - \frac{\partial f}{\partial y}(x_0, y_0)^{-1}(f(x, y));$$

then, for each x, the solutions of the equation $f(x, y) = 0$ are exactly the fixed points of $g(x, \cdot)$. Observe that g is of class C^k and $\frac{\partial g}{\partial y}(x_0, y_0) = 0$, so shrinking U and V if necessary we may suppose that $\|\frac{\partial g}{\partial y}(x, y)\| \le \alpha < 1$ for all x, y.

Let $c = \|\frac{\partial f}{\partial y}(x_0, y_0)^{-1}\|$, fix $\varepsilon < \frac{1}{c}$, and write

$$f(x, y) - f(x_0, y_0) = \frac{\partial f}{\partial x}(x_0, y_0)(x - x_0) + \frac{\partial f}{\partial y}(x_0, y_0)(y - y_0) + R(x, y).$$

Next, fix $\delta_x, \delta_y > 0$ such that $B_{\delta_x}(x_0) \times \overline{B_{\delta_y}(y_0)} \subset U \times V$ and

$$\|R(x, y)\| \le \varepsilon(\|x - x_0\| + \|y - y_0\|)$$

for $\|x - x_0\| < \delta_x$ and $\|y - y_0\| \le \delta_y$. Then

$$\|g(x, y) - g(x_0, y_0)\| = \left\| \frac{\partial f}{\partial y}(x_0, y_0)^{-1} \left(\frac{\partial f}{\partial x}(x_0, y_0)(x - x_0) + R(x, y) \right) \right\|$$

$$\le ck\|x - x_0\| + c\varepsilon(\|x - x_0\| + \|y - y_0\|),$$

where $k = \|\frac{\partial f}{\partial x}(x_0, y_0)\|$.

Making δ_x smaller if necessary, we may suppose that $c(k + \varepsilon)\delta_x \le (1 - c\varepsilon)\delta_y$. Thus, since $g(x_0, y_0) = y_0$, it follows that $g(B_{\delta_x}(x_0) \times \overline{B_{\delta_y}(y_0)}) \subset \overline{B_{\delta_y}(y_0)}$. Moreover, because $\|\frac{\partial g}{\partial y}(x, y)\| \le \alpha < 1$, we conclude that g is a uniform contraction, and thus there is a unique continuous $\phi : B_{\delta_x}(x_0) \to \overline{B_{\delta_y}(y_0)}$ such that $\phi(x) = g(x, \phi(x))$ for all x. Obviously, $\phi(x_0) = y_0$. The smoothness of ϕ follows from Lemma 2.5 in the appendix. $\square$

As the previous proof suggests, if we are only interested in the existence of the implicit function, regardless of its smoothness properties, then weaker conditions may be assumed. The assumption that f is a C^1 mapping was used to prove that g

is a uniform contraction, but the result is still valid if we are able to prove this latter fact by other means. It is well known, however, that even in the finite-dimensional case one cannot just drop the hypothesis of continuity of the derivatives: thus, one should be careful about the exact meaning of the expression *weaker conditions*.

Also, it is worth noticing that the theorem is equivalent to the inverse function theorem, that is, each of them serves to prove the other one (Exercise 2.10). In Chap. 5 we shall see that both results can be proven for a rather general class of mappings under an appropriate local injectivity assumption, without talking at all about differentiability.

There are many beautiful applications of Theorem 2.5 in the field of boundary value problems. As an elementary example, we may take once again the Dirichlet problem for a second-order equation, but now with a parameter λ:

$$u''(t) = \lambda f(t, u(t)), \qquad u(0) = u(1) = 0. \tag{2.3}$$

If we assume that $f : [0,1] \times \mathbb{R} \to \mathbb{R}$ is continuous and continuously differentiable in its second variable, then (2.3) has a solution when $|\lambda|$ is small enough. Indeed, it suffices to look for zeros of the mapping $F(\lambda, u) = u'' - \lambda f(\cdot, u)$ over some appropriate Banach space. Because $F(0,0) = 0$, we might try to apply the implicit function theorem in order to find a neighborhood of $0 \in \mathbb{R}$ and a continuum $u(\lambda)$ of solutions of (2.3). It becomes quite clear that, in our setting, it is reasonable to consider $X := \mathbb{R}$, but what about Y and Z? From the definition of F we need twice differentiable functions; from the Dirichlet conditions, solutions vanish at the boundary. So, as mentioned in this chapter's introduction, the space

$$Y := \{u \in C^2([0,1]) : u(0) = u(1) = 0\}$$

seems to be a good choice. After taking second-order derivatives, C^2 functions become continuous, so we shall simply take $Z = C([0,1])$. Now we must verify that all the hypotheses are satisfied.

In the first place, F is well defined and (exercise!) continuous. But we need to prove continuous differentiability, so let us compute the partial derivatives. In the first place, we have

$$\frac{\partial F}{\partial \lambda}(\lambda, u)(h) = \lim_{s \to 0} \frac{F(\lambda + sh, u) - F(\lambda, u)}{s} = -hf(\cdot, u).$$

Here, h is just a real number; the result—we must confess—is not too surprising since F is a linear function with respect to λ. The remaining derivative is more interesting:

$$\frac{\partial F}{\partial u}(\lambda, u)(k) = \lim_{s \to 0} \frac{F(\lambda, u + sk) - F(\lambda, u)}{s} = k'' - \lambda \lim_{s \to 0} \frac{f(\cdot, u + sk) - f(\cdot, u)}{s}.$$

Thus, using, for example, Lagrange's theorem (the standard one) we obtain the following linear operator:

$$\frac{\partial F}{\partial u}(\lambda, u)(k) = k'' - \lambda \frac{\partial f}{\partial u}(\cdot, u)k.$$

It is important to observe that we have not proven differentiability yet, but we have enough evidence to make a "guess": $DF(\lambda, u)(h, k) = -hf(\cdot, u) + k'' - \frac{\partial f}{\partial u}k$. Next, one must verify that

$$\lim_{\|(h,k)\| \to 0} \frac{F(\lambda + h, u + k) - F(\lambda, u) - DF(\lambda, u)(h, k)}{\|(h, k)\|} = 0.$$

This task is left to the reader (Exercise 2.9 may help). Fortunately, after the dirty job has been done, the continuity of the differential $DF : X \times Y \to L(X \times Y, Z)$ follows immediately from the previous formula for $DF(\lambda, u)$.

Besides all these details, now comes the best part. We just need to check that $\frac{\partial F}{\partial u}(0, 0)$ is invertible: in this case, from the implicit function theorem there is an open interval I containing 0 and a (unique) C^1 function $u : I \to Y$ that gives, for each λ, a solution of the problem. But the linear operator $L := \frac{\partial F}{\partial u}(0, 0) : Y \to Z$ is simply $Lu := u''$, which is (as "implicitly" mentioned before in this chapter) one-to-one: in other words, for each continuous function $\varphi \in C([0, 1])$ there exists a unique solution u of the Dirichlet problem

$$u'' = \varphi, \qquad u(0) = u(1) = 0.$$

We already know that L is a continuous operator, so it remains to prove the continuity of L^{-1}. And this is a very well-known fact: the reader who is familiar with functional analysis may invoke the open mapping theorem or the Fredholm alternative. But it is also true that, in this case, the proof is straightforward. Because L^{-1} is linear, it suffices to prove the continuity at the origin, so let $\varphi_n \to 0$ and $u_n = L^{-1}\varphi_n$. Then $\|u_n''\|_\infty = \|\varphi_n\|_\infty \to 0$ and, using the bounds in Appendix B.2.2, $\|u_n\|_\infty \le \frac{1}{8}\|u_n''\|_\infty \to 0$ and $\|u_n'\|_\infty \le \|u_n''\|_\infty \to 0$. We conclude that $\|u_n\|_{C^2} \to 0$, and the proof is complete.

Remark 2.2. The preceding proof is stronger than the one in Appendix B.2, Exercise 5, where continuity of $L^{-1} : Z \to Y$ was proven with Y regarded as a subspace of Z (namely, with the uniform norm).

After such a successful experience, we may take a second look at Theorem 2.5 and ask: do the previous computations imply that solutions of (2.3) are unique when $|\lambda|$ is small? Of course, this is trivial if $\lambda = 0$, but the answer is, in general, negative. The implicit function theorem ensures the existence of a unique C^1 graph $(\lambda, u(\lambda))$ passing trough $(0, 0)$, but nothing prevents the existence of "large" solutions, different from $u(\lambda)$, for some values of λ.

Now, what happens if we know in advance that solutions are unique? In this case, an existence result seems, at first sight, a bit tautological: if we say that the solution is unique, before anything else we are assuming that a solution exists. Of course, this is not a satisfactory conclusion since what we really mean by *uniqueness* is that the problem has *at most* one solution. But is this enough to prove existence? As we shall see, in many cases the answer is affirmative. This is hidden, for example, in the aforementioned Fredholm alternative or, to give an even more trivial example, in the dimension theorem for linear operators: for the equation $Ax = b$ with $A \in \mathbb{R}^{n \times n}$ and $b \in \mathbb{R}^{1 \times n}$, it is more than evident that "uniqueness implies existence."

So let us conduct a quick review of what we have done in all the preceding pages: do we already know of any uniqueness results? Let's think... yes, we shall talk, once again, about the Dirichlet problem for the second-order problem with nondecreasing f. In this case, assuming that f is a C^1 mapping, it is not difficult to explain, through the implicit function theorem, why the problem always has a solution. First, add a parameter λ to the problem to obtain (2.3). We proved that the problem has solutions when λ is small, but in this case, the same method provides a much better result. Indeed, if u_0 is a solution for some $\lambda_0 \geq 0$, then it is not difficult to prove (using monotonicity) that $\frac{\partial F}{\partial u}(\lambda_0, u_0)$ is also invertible; thus, the implicit function theorem ensures the existence of solutions for λ in a neighborhood of λ_0. In principle, this does not imply that the branch of solutions $u(\lambda)$ can be continued up to $\lambda = 1$, but a direct argument shows that the set $\{\lambda \in [0,1] : (2.3) \text{ has a solution}\}$ is also closed. Hence, a (unique) solution exists for every value of λ. More details are given in Exercise 2.8.

We end this chapter with an alternative proof of the Picard existence-uniqueness theorem for ordinary differential equations. In fact, we shall prove that the result is valid under stronger assumptions, but the classical case follows from a slightly more general version of the implicit function theorem.

For simplicity, assume that $f : \mathbb{R} \times \mathbb{R}^n \to \mathbb{R}^n$ is a C^1 function and that $t_0 = 0$. We want to solve the problem

$$x'(t) = f(t, x(t)), \qquad x(0) = x_0.$$

Setting $y(t) := x(\lambda t)$ for some $\lambda > 0$, the problem becomes

$$y'(t) = \lambda f(\lambda t, y(t)), \qquad y(0) = x_0. \tag{2.4}$$

Hence, it suffices to prove that (2.4) has a solution for λ small enough. Let $X = \mathbb{R}$, $Y := \{y \in C^1([0,1], \mathbb{R}^n) : y(0) = x_0\}$, and $Z = C([0,1], \mathbb{R}^n)$. It is easy to check that the function $F : \mathbb{R} \times Y \to Z$ given by $F(\lambda, y)(t) := y'(t) - \lambda f(\lambda t, y((t)))$ is well defined and of class C^1. Moreover, $\frac{\partial F}{\partial y}(0,0)(\varphi) = \varphi'$, which is an invertible operator with inverse $\psi \mapsto x_0 + \int_0^t \psi(s)\,ds$, so the result follows. The same conclusion is obtained if we consider instead $F : \mathbb{R} \times C([0,1], \mathbb{R}^n) \to C([0,1], \mathbb{R}^n)$ given by

$$F(\lambda, y)(t) := y(t) - x_0 - \lambda \int_0^t f(\lambda s, y((s)))\,ds.$$

Incidentally, we may observe two extra facts. On the one hand, from the regularity part of Theorem 2.5 (coming soon...), it is easy to prove at once that if f is a C^k mapping then the solution is also C^k as a function of the initial data x_0. On the other hand, the preceding proof is also valid for ordinary differential equations in Banach spaces. Details are, once again, left to the reader.

Appendix

This appendix is devoted to proving the smoothness of the implicit function obtained in Theorem 2.5. At this stage, we only need to prove that if f is a C^k uniform contraction, then the fixed point $y = y(x)$ of $f(x, \cdot)$ is also a C^k mapping. Let us firstly recall a basic fact.

Lemma 2.4. *Let X be a Banach space and let $T \in L(X, X)$ satisfy $\|T\| < 1$. Then $I - T$ is invertible and $\|(I - T)^{-1}\| \leq \frac{1}{1 - \|T\|}$.*

To check this, one may take inspiration from the sum of the geometric series and verify that

$$(I - T)^{-1}(x) = \sum_{n=0}^{\infty} T^n x.$$

Alternatively, just apply the contraction mapping theorem to obtain, for each x, a fixed point $y = y(x)$ of the equation $y = Ty + x$. No matter which of the two approaches is employed, it is immediately seen that $\|(I - T)^{-1}\| \leq \frac{1}{1 - \|T\|}$.

We are now in a position to prove the crucial lemma.

Lemma 2.5. *Let $f : U \times \overline{V} \to \overline{V}$ be a C^k mapping, where U and V are respectively subsets of some Banach spaces X and Y. If f is a uniform contraction, then the function $y : U \to \overline{V}$ defined implicitly from the equation $y = f(x, y)$ is a C^k mapping.*

Proof. We shall prove only the case $k = 1$; the induction step is left as an exercise. Let us start, once again, with a naïve approach. From the equality $y(x) = f(x, y(x))$ it is deduced that the differential of y (*if it exists*) must satisfy the equality

$$Dy(x) = \frac{\partial f}{\partial x}(x, y(x)) + \frac{\partial f}{\partial y}(x, y(x)) \circ Dy(x)$$

or, equivalently,

$$\left[I - \frac{\partial f}{\partial y}(x, y(x))\right] \circ Dy(x) = \frac{\partial f}{\partial x}(x, y(x)).$$

Because $\|\frac{\partial f}{\partial y}(x, y)\| \leq \alpha < 1$ for all x, y, the operator in square brackets is invertible, and hence

$$Dy(x) = \left[I - \frac{\partial f}{\partial y}(x, y(x))\right]^{-1} \circ \frac{\partial f}{\partial x}(x, y(x)).$$

This looks rather awful, but the right-hand-side term is a well-defined operator and depends continuously on x, so it suffices to check that the "remainder"

$$R(x + h) := y(x + h) - y(x) - Dy(x)(h)$$

satisfies $\frac{R(x+h)}{\|h\|} \to 0$ as $h \to 0$. Because x is fixed, we may set $y = y(x)$ and $k = y(x + h) - y$ and use the formula

$$f(x+h,y+k) - f(x,y) = \frac{\partial f}{\partial x}(x,y)(h) + \frac{\partial f}{\partial y}(x,y)(k) + R_f(x+h,y+k).$$

Now let us observe two important facts: first, the left-hand side is equal to $y(x+h) - y = k$. Second, it follows from the definition of R that

$$\left[I - \frac{\partial f}{\partial y}(x,y)\right](R(x+h)) = \left[I - \frac{\partial f}{\partial y}(x,y)\right](k) - \frac{\partial f}{\partial x}(x,y)(h)$$

$$= k - \frac{\partial f}{\partial y}(x,y)(k) - \frac{\partial f}{\partial x}(x,y)(h) = R_f(x+h,y+k)$$

so, because $I - \frac{\partial f}{\partial y}(x,y)$ is invertible, we only need to prove that

$$\lim_{h \to 0} \frac{R_f(x+h,y+k)}{\|h\|} = 0.$$

We know that $\frac{R_f(x+h,y+k)}{\|(h,k)\|} \to 0$ as $\|(h,k)\| \to 0$, so it suffices to prove that $\frac{\|k\|}{\|h\|}$ remains bounded as $\|h\| \to 0$. But this is clear since

$$\|k\| = \|y(x+h) - y\| = \|f(x+h,y+k) - f(x,y)\|$$

$$\leq \|f(x+h,y+k) - f(x+h,y)\| + \|f(x+h,y) - f(x,y)\|$$

$$\leq \alpha\|k\| + \left\|\frac{\partial f}{\partial x}(x,y)(h) + R_f(x+h,y)\right\|.$$

Then

$$(1-\alpha)\frac{\|k\|}{\|h\|} \leq \left\|\frac{\partial f}{\partial x}(x,y)\right\| + \frac{\|R_f(x+h,y)\|}{\|h\|},$$

and the result follows. □

Problems

2.1. Find an example of a complete metric space X and a function $T : X \to X$ with no fixed points such that

$$d(Tx,Ty) < d(x,y) \qquad \text{for all } x,y \in X, x \neq y.$$

Is it possible to choose X compact?

Note: it can be proven that if X is a closed bounded convex subset of a *uniformly convex* Banach space, then T has a necessarily (unique) fixed point. Furthermore, existence follows under the weaker assumption that T is *nonexpansive*, that is, $d(Tx,Ty) \leq d(x,y)$ for all $x,y \in X$ (see [21] for details).

2.2. Prove that if $f : [0,1] \times \mathbb{R}^{2n} \to \mathbb{R}^n$ is continuous and Lipschitz with respect to (u,u') with constant L small enough, then the system

$$\begin{cases} u''(t) - a(t)u(t) = f(t,u(t),u'(t)) \\ u(0) = u_0, \quad u(1) = u_1 \end{cases}$$

has a unique solution for any continuous $a : [0,T] \to \mathbb{R}_{\geq 0}$ and $u_0, u_1 \in \mathbb{R}^n$. Generalize for Neumann and periodic boundary conditions, assuming that $a \neq 0$.

2.3. 1. Formulate and prove an existence-uniqueness result for the delay differential equation

$$x'(t) = f(t,x(t),x(t-\tau)), \tag{2.5}$$

where $f : \mathbb{R} \times \mathbb{R}^{2n} \to \mathbb{R}^n$, and $\tau > 0$ is a given constant, with continuous initial condition

$$x(t) = \varphi(t) \quad \text{for } t \in [-\tau,0]. \tag{2.6}$$

2. Prove that if $f = f(t,x,y)$ is continuous, T-periodic in its first coordinate for some $T > \tau$ and globally Lipschitz in (x,y) with sufficiently small Lipschitz constant, then (2.5) admits at least one T-periodic solution.
   Hint: define the Poincaré-type operator $\varphi \mapsto x|_{[T-\tau,T]}$, where x is the unique solution of the initial value problem obtained in 1.

2.4. Define a Picard successive approximations scheme for the preceding two exercises.

2.5. Prove the *Peano theorem*: if $f : \Omega \subset \mathbb{R} \times \mathbb{R}^n \to \mathbb{R}^n$ is continuous and $(t_0,x_0) \in \Omega$, then the problem

$$\begin{cases} x'(t) = f(t,x(t)) \\ x(t_0) = x_0 \end{cases} \tag{2.7}$$

has at least one solution defined in a neighborhood of t_0.
   Hint: fix $\hat{\delta}, r > 0$ such that $K := [t_0 - \hat{\delta}, t_0 + \hat{\delta}] \times \overline{B_r(x_0)} \subset \Omega$. Using the Stone–Weierstrass theorem, consider C^1 functions $f_n : \mathbb{R} \times \mathbb{R}^n \to \mathbb{R}^n$ such that $f_n|_K \to f|_K$ uniformly and $\|f_n\|_\infty \leq M$ for some constant M independent of n. Let $x_n : \mathbb{R} \to \mathbb{R}^n$ be the unique solution of the problem

$$\begin{cases} x_n'(t) = f_n(t,x_n(t)) \\ x_n(t_0) = x_0, \end{cases}$$

and fix $\delta = \min\{\hat{\delta}, \frac{r}{M}\}$. Using the Arzelá–Ascoli theorem, deduce the existence of a subsequence $\{x_{n_j}\}$ that converges uniformly on $[t_0 - \delta, t_0 + \delta]$ to a solution of (2.7).

2.6. * (Adapted from [66]) Let $t_0 \in \mathbb{R}$, $x_0 \in \mathbb{R}^n$ and $f : [t_0, t_0 + A] \times \overline{B_R(x_0)} \to \mathbb{R}^n$ continuous. Set $M := \|f\|_\infty$. Assume there exists $g : [t_0, t_0 + A] \times [0, 2R] \to [0, M]$ continuous and nondecreasing such that

$$|f(t,x) - f(t,y)| \leq g(t, |x-y|)$$

for $t_0 \le t \le t_0 + A$, $x, y \in \overline{B_R(x_0)}$. Further, assume that $g(\cdot, 0) \equiv 0$ and that the problem

$$u'(t) = g(t, u(t)), \qquad u(t_0) = 0$$

does not admit nontrivial solutions. Prove that the successive approximations scheme defined by

$$x_{n+1}(t) := x_0 + \int_{t_0}^{t} f(s, x_n(s)) \, ds$$

converges uniformly on $[t_0, t_0 + \delta]$ to a (unique) solution of the problem

$$x'(t) = f(t, x(t)), \qquad x(t_0) = x_0$$

with $\delta = \min\{A, \frac{R}{M}\}$.

2.7. Let E and F be normed spaces, and let $f : E \to F$ be a C^1 function.

1. Prove (without using the Hahn–Banach theorem!) that for every $x, y \in E$,

$$\|f(x) - f(y)\| \le M\|x - y\|,$$

where $M := \max_{t \in [0,1]} \|Df(tx + (1-t)y)\|$.
Hint: given $\varepsilon > 0$, define the function

$$\varphi(t) := \|f(tx + (1-t)y) - f(y)\| - (M + \varepsilon)t\|x - y\|$$

and verify that $\sup\{t \in [0,1] : \varphi(t) \le 0\} = 1$.
2. Deduce that if E is complete, $F = E$, and $\|Df(x)\| \le \alpha < 1$ for all x, then f has a unique fixed point.

2.8. Let $f : [0,1] \times \mathbb{R}^n \to \mathbb{R}^n$ be continuous, locally Lipschitz, and nondecreasing in u, that is,

$$[f(t, u) - f(t, v)] \cdot (u - v) \ge 0$$

for all $t \in [0,1]$, $u, v \in \mathbb{R}^n$. For $w \in L^2((0,1), \mathbb{R}^n)$, let us define u_w as the unique solution of the problem

$$u''(t) = f(t, W(t)), \qquad u(0) = u(1) = 0,$$

where $W(t) := \int_0^t w(s) \, ds$. Finally, consider

$$H := \{w \in L^2((0,1), \mathbb{R}^n) : \int_0^1 w(t) \, dt = 0\}$$

and the operator $T : H \to H$ given by $Tw := u_w'$.

1. Prove that T is well defined and monotone nonincreasing.
2. Deduce that the Dirichlet problem

$$u''(t) = f(t, u(t)), \qquad u(0) = u(1) = 0$$

has a unique solution.

3. Show that if f is also continuously differentiable with respect to u, then the result can be deduced from the implicit function theorem.

   Hint: add a parameter $\lambda \geq 0$ to the equation and consider the set $\mathscr{A}$ of those values of λ such that the problem admits a solution. From the implicit function theorem, it is verified that $\mathscr{A}$ is open in $[0, +\infty)$. Using a priori bounds and the Arzelá–Ascoli theorem, prove that $\mathscr{A}$ is also closed.

2.9. Let X, Y, and Z be normed spaces, and let $f : U \times V \to Z$ be continuous, where $U \subset X$ and $V \subset Y$ are open. Assume that for all $x \in U$ and $y \in V$ the functions $f(\cdot, y)$ and $f(x, \cdot)$ are differentiable, and define $\frac{\partial f}{\partial x}(x, y) \in L(X, Z)$ and $\frac{\partial f}{\partial y}(x, y) \in L(Y, Z)$ by

$$\frac{\partial f}{\partial x}(x, y) = Df(\cdot, y)(x), \qquad \frac{\partial f}{\partial y}(x, y) = Df(x, \cdot)(y).$$

Prove that "C^1 implies C^1," that is, if $\frac{\partial f}{\partial x} : U \times V \to L(X, Z)$ and $\frac{\partial f}{\partial y} : U \times V \to L(Y, Z)$ are continuous, then f is a C^1 function. Is this a tautology?

2.10. 1. Using the implicit function theorem, prove the *inverse function theorem*: let X, Y be Banach spaces, and let $U \subset X$ be open. If $F : U \to Y$ is C^1 and $DF(x_0) : X \to Y$ is an isomorphism for some $x_0 \in U$, then there exist U_0, V_0 neighborhoods of x_0 and $F(x_0)$, respectively, such that $F : U_0 \to V_0$ is a C^1 diffeomorphism. Furthermore, $DF^{-1}(F(x)) = DF(x)^{-1}$ for all $x \in U_0$.
2. Deduce the implicit function theorem from the inverse function theorem.

2.11. (Lagrange multipliers) Let $F, G : U \to \mathbb{R}$ be C^1 functions, where U is an open subset of a Banach space, and let x_0 be a critical point of F subject to the condition $G(x) = 0$. If $DG(x_0) \neq 0$, then there exists $\lambda \in \mathbb{R}$ such that $DF(x_0) = \lambda DG(x_0)$.

2.12. Consider the problem

$$u''(t) + \lambda e^{u(t)} = 0, \qquad u(0) = u(1) = 0,$$

with $\lambda > 0$.

1. Using the implicit function theorem, prove the existence of $\lambda_0 > 0$ such that the problem has a solution for $0 < \lambda < \lambda_0$. Why is this result immediate when $\lambda \leq 0$?
2. Prove that all solutions are positive in $(0, 1)$.
3. Prove the existence of $\lambda^* > \lambda_* > 0$ such that the problem has two solutions for $\lambda \in (0, \lambda_*)$ and no solutions for $\lambda > \lambda^*$.

   Hint: multiply the equation by u' and integrate. Observe that u is symmetric and achieves a unique absolute maximum M at $t = \frac{1}{2}$, with

$$\int_0^M \frac{dv}{\sqrt{e^M - e^v}} = \frac{\sqrt{2\lambda}}{2}. \tag{2.8}$$

Conversely, prove that for each M that verifies (2.8) there exists a solution u with $\|u\|_\infty = M$. Now draw the left-hand-side term of (2.8) as a function of M.

2.13. Let $a \notin \mathbb{N}_0^2$. Prove that the forced pendulum equation

$$u''(t) + a \sin u(t) = p(t)$$

admits 2π-periodic solutions for any $p : \mathbb{R} \to \mathbb{R}$ continuous and 2π-periodic such that $\|p\|_\infty$ is small enough.

Hint: set appropriate spaces and apply the implicit function theorem to the mapping $F(u, p) := u'' + a \sin u - p$.

Chapter 3
Iterative Methods

3.1 Monotone Iterations: Upper and Lower Solutions

3.1.1 Warming Up Engines: The Scalar Case

Let us begin with a very simple example: suppose we want to solve the fixed point equation

$$x = f(x),$$

where $f : \mathbb{R} \to \mathbb{R}$ is a continuous function. We have few ideas in mind today, so we start just by trying out a single value $\alpha \in \mathbb{R}$ to obtain, say, $\alpha < f(\alpha)$.

This inequality shows, first, that we have not solved the problem yet—bad luck! But this is not enough to lose hope: let us call α a (strict) *lower solution* of the problem and observe that if we also have an *upper solution*, that is, a value β such that $\beta \geq f(\beta)$, then a solution between α and β necessarily exists. This is a trivial consequence of Bolzano's theorem applied to the function $g(x) = f(x) - x$; in particular, the same theorem allows us to obtain such a solution by iteration using the most elementary of all iterative schemes: the bisection method. However, we are looking for a method that could be extended to much more general situations; with this aim, we shall introduce the idea of *monotone iterations*.

Let us assume for a moment that f is nondecreasing: in this case, if we start at $u_0 := \alpha$ and define the iteration $u_{n+1} := f(u_n)$, then it is readily verified that $\{u_n\}$ is a nondecreasing sequence of lower solutions. Indeed:

$$u_1 = f(\alpha) > \alpha,$$

$$u_2 = f(u_1) \geq f(\alpha) = u_1,$$

$$u_3 = f(u_2) \geq f(u_1) = u_2,$$

and so on. Certainly, this does not suffice to ensure its convergence, unless we know in advance that the sequence is bounded. And this is very easy to prove if our couple (α, β) is "well ordered," namely $\alpha \leq \beta$. In this case,

P. Amster, *Topological Methods in the Study of Boundary Value Problems*, Universitext, DOI 10.1007/978-1-4614-8893-4_3, © Springer Science+Business Media New York 2014

$$u_1 = f(\alpha) \le f(\beta) \le \beta,$$

$$u_2 = f(u_1) \le f(\beta) \le \beta,$$

and inductively it follows that $u_n \le \beta$ for all n. Thus, the sequence converges to some u; moreover, since $f(u_n) = u_{n+1} \to u$, we conclude that u is a solution of the fixed point equation. If we start at β instead of α, then the result is a nonincreasing sequence of upper solutions, which also converges to a (possibly different) solution.

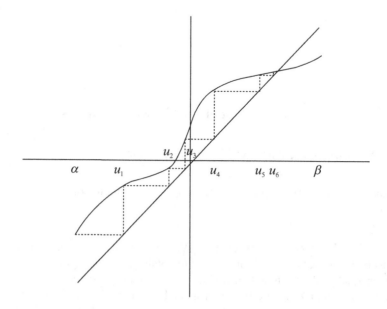

Note that the preceding sequences may not converge if we do not impose the monotonicity assumption on f; however, we know that at least one fixed point of f in $[\alpha, \beta]$ exists. Is there any way to obtain it using monotone iterations?

To provide an answer to this question, assume that f is an arbitrary C^1 function and that $\alpha \le \beta$ are respectively a lower and an upper solution. The key idea is simple: we are looking for solutions between α and β, so we may just forget about the behavior of f *outside* the interval $[\alpha, \beta]$. All our problems started when we relaxed the assumption that f is nondecreasing; nonetheless, if we consider only its restriction to $[\alpha, \beta]$, then it is very easy to overcome this difficulty. Indeed, for a sufficiently large constant λ, the function $f(x) + \lambda x$ is nondecreasing over $[\alpha, \beta]$: it is enough to take, for example, $\lambda = -\min_{x \in [\alpha, \beta]} f'(x)$. Next, observe that the fixed point problem $x = f(x)$ is equivalent to the equation

$$x + \lambda x = f(x) + \lambda x,$$

which may also be written as a fixed point problem:

$$x = \frac{f(x) + \lambda x}{1 + \lambda} := g(x).$$

Now $g|_{[\alpha,\beta]}$ is a nondecreasing function, and it is clear that $\alpha \leq g(\alpha)$ and $\beta \geq g(\beta)$, so we are again in the monotone case. When f is just continuous, such a λ may not exist but, as we learned in previous chapters, one can always use an approximation argument.

The situation is different when $\alpha \geq \beta$: here, the sequences do not converge unless more solutions exist outside the interval $[\alpha, \beta]$.

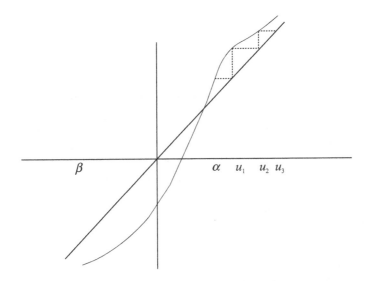

The figure below suggests that the fixed point of f can be obtained in this case by an *inverse* process. Graphically, this only means that now we join the point $(\alpha, f(\alpha))$ with the diagonal through a *vertical* segment, move horizontally to the left until we meet the graph of f, and then repeat the process again and again.

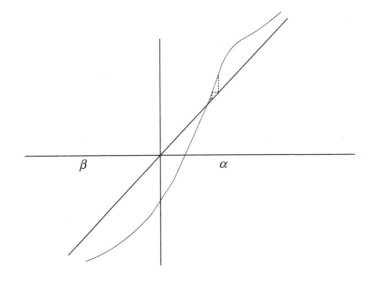

Of course, this is not a big surprise: in this very elementary case, we may assume that f is nondecreasing; then, setting $\gamma := f(\alpha)$ and $\delta = f(\beta)$, the process can be reproduced for the inverse function $f^{-1} : [\delta, \gamma] \to [\beta, \alpha]$. Observe that $\gamma = f(\alpha) \geq \alpha = f^{-1}(\gamma)$ and $\delta = f(\beta) \leq \beta = f^{-1}(\delta)$; thus, (δ, γ) is a well-ordered couple of a lower and an upper solution for the fixed point equation $y = f^{-1}(y)$. The iteration $y_{n+1} = f^{-1}(y_n)$, starting at $y_0 = \delta$ or $y_0 = \gamma$, converges to a fixed point of f^{-1}, which is obviously also a fixed point of f.

3.1.2 Monotone Iterations in Abstract Spaces

Those readers who did not skip the previous chapters must already be convinced that fixed point methods are quite important for our purposes. Shooting methods use mainly Brouwer-type theorems, but, more generally, we have already seen that many boundary value problems can be studied as infinite-dimensional fixed point equations. Thus, a natural question arises: how can the method of monotone iterations be generalized in such a way that it allows us to find fixed points of operators in Banach spaces E?

First, we need to introduce in E an *order*, which must be compatible with the vector space structure. In more precise words, we shall consider a reflexive, antisymmetric, and transitive relation $\leq$ in E, which also satisfies the following conditions:

- If $x \leq y$, then $x + z \leq y + z$ for all $z \in E$.
- If $x \leq y$, then $cx \leq cy$ for all $c \in R_{\geq 0}$.

In this case, the set $K := \{x \in E : x \geq 0\}$ of "nonnegative" elements of E has the following properties:

1. $K + K \subset K$.
2. $K \cap -K = \{0\}$.
3. K is convex.

Indeed, the first property is an immediate consequence of the first compatibility condition. The second one follows from the fact that 0 is clearly an element of K and, moreover, if $u \in K \cap -K$, then $u \geq 0$ and $-u \geq 0$. Adding u to the last inequality, it is seen that $u \leq 0$, and hence $u = 0$. Finally, 3 is deduced from 1 and the second compatibility condition: if $x, y \geq 0$ and $t \in [0, 1]$, then $tx \geq 0$ and $(1 - t)y \geq 0$, which implies $tx + (1 - t)y \geq 0$.

Conversely, any arbitrary set $K \subset E$ satisfying 1, 2, and 3 induces a compatible partial order given by

$$x \leq y \text{ iff } y - x \in K.$$

Reflexivity is obvious since $0 \in K$; antisymmetry is a clear consequence of 2: if $y - x \in K$ and $x - y \in K$, then $x - y = 0$. Transitivity is due to 1: if $x \leq y$ and $y \leq z$, then $y - x \in K$ and $z - y \in K$, so $z - x = z - y + y - x \in K$.Regarding the compatibility properties, let $x \leq y$; then $(y+z) - (x+z) = y - x \in K$, so $x+z \leq y+z$. On the other hand, using 1 it is proven inductively that $nK \subset K$ for every $n \in \mathbb{N}$, and by convexity (and Archimedes principle!) we conclude that $cK \subset K$ for all $c \geq 0$.

A subset K of a vector E satisfying 1–3 is called a *cone*; intuitively, this denomination is understood from the fact that K is a nonempty convex set that consists of rays emanating from the origin. Condition 2 ensures that K contains only *half*-lines and never whole lines.

If E is a normed space, then we shall also require compatibility with the topology, more precisely: if $x_n \to x$, $y_n \to y$, and $x_n \leq y_n$, then $x \leq y$. But this is equivalent to the following property:

$$\text{If } z_n \geq 0 \text{ and } z_n \to z, \text{ then } z \geq 0,$$

so we are just saying that the order induced by K is compatible with the norm if and only if K is closed. An important consequence of compatibility is the fact that if $\{x_n\}$ is a nondecreasing sequence and converges to some x, then $x_n \leq x$ for all n: indeed, for $m \geq n$ we know that $x_m - x_n \geq 0$, and hence $x - x_n = \lim_{m\to\infty} x_m - x_n \geq 0$.

Elementary examples of compatible cones include

$$K = \{x \in \mathbb{R}^n : x_j \geq 0 \text{ for } j = 1,\ldots,n\},$$

which induces the standard coordinate-by-coordinate order in $\mathbb{R}^n$, or

$$K = \{u \in C([0,1]) : u(t) \geq 0 \text{ for all } t \in [0,1]\},$$

which induces the pointwise order in $C([0,1])$.

It is time to try and see if the method of monotone iterations works in the abstract context. Let E be a Banach space, equipped with an order $\leq$ induced by some closed cone K, and let $T : E \to E$ be a continuous monotone nondecreasing operator, that is: $x \leq y \Rightarrow Tx \leq Ty$. As before, we shall say that α and β are a lower and an upper solution of the fixed point problem $x = Tx$ if

$$\alpha \leq T\alpha \qquad \text{and} \qquad \beta \geq T\beta,$$

respectively. We shall say that α and β are well ordered if $\alpha \leq \beta$; in this case, we have, as previously, sequences

$$\alpha = u_0 \leq u_1 \leq u_2 \leq \ldots \leq \beta,$$

$$\beta = v_0 \geq v_1 \geq v_2 \geq \ldots \geq \alpha,$$

where u_n and v_n are recursively defined by $u_{n+1} = Tu_n$ and $v_{n+1} = Tv_n$. Now an obvious question arises: is there anything that might guarantee the convergence of these sequences?

As even the most optimistic one can imagine, the answer to this question is, in general, negative. In the first place, if E is infinite-dimensional, then there exist bounded sequences without convergent subsequences; in the second place, to be bounded in the sense of the order $\leq$ does not imply, in general, to be bounded in the sense of the norm. Let us fix this last hole first.

Definition 3.1. Let E be a Banach space and $K \subset E$ a closed cone. We shall say that the order $\leq$ induced by K is *normal* if there exists a constant $c > 0$ such that

$$0 \leq x \leq y \Rightarrow \|x\| \leq c\|y\|$$

for all x, y.

For example, the usual order of $C([0,1])$ is normal with $c = 1$: if $0 \leq u(t) \leq v(t)$ for all $t \in [0,1]$, then $\|u\|_\infty \leq \|v\|_\infty$.

Normality guarantees that the previous sequences of lower and upper solutions are bounded: indeed, for the first sequence we have, for all n,

$$0 \leq u_n - \alpha \leq \beta - \alpha.$$

Hence $\|u_n - \alpha\| \leq c\|\beta - \alpha\|$, which in turn implies

$$\|u_n\| \leq \|\alpha\| + c\|\beta - \alpha\|.$$

The proof for $\{v_n\}$ is analogous.

As previously mentioned, this is not enough to ensure convergence, but an extra hypothesis on the continuous operator T will help.

Definition 3.2. Let X and Y be normed spaces, and let $U \subset X$. A continuous operator $T : U \to Y$ is called *compact* if $\overline{T(B)}$ is compact for every bounded set $B \subset U$.

This property is usually satisfied by the operators we shall be dealing with and proves to be extremely useful when applying fixed point methods. In the present context, because we already know that the sequence $\{u_n\}$ is bounded, this implies that the closure of the set $\{Tu_n : n \geq 1\} = \{u_n : n \geq 2\}$ is compact. Thus, there exists a subsequence u_{n_j} that converges monotonically to some u; in particular, this implies that $u_{n_j} \leq u$ for all j. Furthermore, for each n we may take j such that $n_j \geq n$; then $u_n \leq u_{n_j} \leq u$, and we may apply a sort of "sandwich theorem": for each n, define $j(n)$ as the largest value such that $n_{j(n)} \leq n$. Hence,

$$u_{n_{j(n)}} \leq u_n \leq u,$$

and taking the limit on the left-hand side of the inequalities we deduce that $u_n \to u$. Formally, for $\varepsilon > 0$ fix j_0 such that $\|u - u_{n_{j_0}}\| < \frac{\varepsilon}{c}$. If $n \geq n_{j_0}$, then

$$0 \leq u - u_n \leq u - u_{n_{j_0}},$$

and by normality

$$\|u - u_n\| \leq c\|u - u_{n_{j_0}}\| < \varepsilon.$$

Summarizing, we have proven the following theorem.

Theorem 3.1. *Let E be a Banach space and $K \subset E$ a closed cone. Assume that the order induced by K is normal and that $T : E \to E$ is compact and nondecreasing. If (α, β) is a well-ordered couple of a lower and an upper solution, then the sequences defined by*

$$u_0 := \alpha, \qquad u_{n+1} = Tu_n$$

$$v_0 := \beta, \qquad v_{n+1} = Tv_n$$

converge respectively to fixed points u, v of T such that $\alpha \le u \le v \le \beta$.

3.1.3 Dirichlet Revisited

Let us consider, once again, the Dirichlet problem

$$u''(t) = f(t, u(t)), \qquad u(0) = u(1) = 0. \tag{3.1}$$

As mentioned in the previous chapter, this problem can be written as the functional equation $Lu = Nu$, where $Lu := u''$ and $Nu := f(\cdot, u)$; moreover, if we set $X = \{u \in C^2([0, 1]) : u(0) = u(1) = 0\}$, then, as shown in the preceding chapter, $L : X \to C([0, 1])$ is invertible and the problem is reduced to the fixed point equation $u = L^{-1}Nu$.

Now let $E = C([0, 1])$ be endowed with the standard pointwise order, and assume that $\alpha, \beta \in E$ are respectively a lower and an upper solution of the problem, that is,

$$\alpha \le L^{-1}N\alpha \qquad \text{and} \qquad \beta \ge L^{-1}N\beta. \tag{3.2}$$

If we suppose, for simplicity, that α and β are C^2 functions, then (3.2) is satisfied, in particular, if the following conditions hold:

$$\alpha''(t) \ge f(t, \alpha(t)), \qquad \beta''(t) \le f(t, \beta(t)) \tag{3.3}$$

for all t, and

$$\alpha(0) \le 0 \le \beta(0), \quad \alpha(1) \le 0 \le \beta(1). \tag{3.4}$$

Condition (3.4) is obviously necessary since $L^{-1}N\alpha$ and $L^{-1}N\beta$ vanish on the boundary; condition (3.3) relies on the fact that L^{-1} is nonincreasing (this can be regarded as a particular case of the maximum principle; see appendix, Sect. B.2.3 for details). Concretely, observe that if $\alpha'' \ge N\alpha$, then $L^{-1}\alpha'' \le L^{-1}N\alpha$ and, by definition, $L^{-1}\alpha'' = \alpha - \varphi$, where $\varphi(t) = (\alpha(1) - \alpha(0))t + \alpha(0) \le 0$ for $t \in [0, 1]$. In the same way, it is seen that $\beta \ge L^{-1}N\beta$. One might get a little confused here since $L^{-1}\alpha''$ and $L^{-1}\beta''$ may be different from α and β, respectively, but this is just because we are not forcing α and β to vanish on the boundary. Condition (3.4) is much more flexible and makes the method more powerful, in the sense that the task of finding α and β becomes easier.

The nonincreasing character of L^{-1} provides a straightforward application of the results in the preceding section. If f is nonincreasing in u, then it is easily verified that the operator $L^{-1}N$ is nondecreasing; thus, if there are functions $\alpha \le \beta$ satisfying (3.3)–(3.4), then problem (3.1) has at least one solution u such that $\alpha \le u \le \beta$. The reader may verify this using Theorem 3.1, after checking that $L^{-1}N$ is compact. However, as we shall be dealing with a more general situation, the details of this particular case are left as an exercise. In any case, compactness will play a crucial role in the following chapters, so we shall return to this issue soon.

Let us try now a direct approach to the general situation. With the one-dimensional case still in mind, we might wish to avoid the monotonicity condition on f, so we might, in the first place, take $\lambda > 0$ and transform the problem into an equivalent one:

$$u''(t) - \lambda u(t) = f(t, u(t)) - \lambda u(t).$$

If we assume that f is a C^1 function with respect to u, then taking λ large enough the function $f(t, u) - \lambda u$ is nonincreasing in u for all t and $u \in [\alpha(t), \beta(t)]$; specifically, it suffices to consider any λ satisfying

$$\lambda \ge \max_{(t,u) \in K} \frac{\partial f}{\partial u}(t, u), \tag{3.5}$$

where $K := \{(t, u) : t \in [0, 1], u \in [\alpha(t), \beta(t)]\}$. As before, since $\lambda > 0$, it is seen that the linear operator $Lu := u'' - \lambda u$ is invertible, and L^{-1} is nonincreasing (appendix, Lemma B.3).

Now we are in a position to provide a direct proof of the method of upper and lower solutions for this particular case.

Theorem 3.2. *Let $f : [0, 1] \times \mathbb{R} \to \mathbb{R}$ be continuous and continuously differentiable with respect to u, and let the functions $\alpha \le \beta$ satisfy*

$$\alpha''(t) \ge f(t, \alpha(t)), \qquad \beta''(t) \le f(t, \beta(t))$$

for all t, and

$$\alpha(0) \le 0 \le \beta(0), \quad \alpha(1) \le 0 \le \beta(1).$$

Then problem (3.1) has at least one solution u such that $\alpha \le u \le \beta$.

Proof. Let $\lambda \ge 0$ satisfy (3.5), and define recursively $u_0 := \alpha$ and u_{n+1} as the unique solution of the linear problem

$$u_{n+1}''(t) - \lambda u_{n+1}(t) = f(t, u_n(t)) - \lambda u_n(t), \qquad u_{n+1}(0) = u_{n+1}(1) = 0. \tag{3.6}$$

In the first place, we claim that u_1 is a lower solution with $\alpha \le u_1 \le \beta$. Indeed, observe that

$$u_1''(t) - \lambda u_1(t) = f(t, \alpha(t)) - \lambda \alpha(t) \le \alpha''(t) - \lambda \alpha(t),$$

and Lemma B.3 applies since $\alpha(0) \leq 0 = u_1(0)$ and $\alpha(1) \leq 0 = u_1(1)$. This shows that $\alpha \leq u_1$. Moreover, as the function $f(t, u) - \lambda u$ is nonincreasing, we obtain

$$u_1''(t) - \lambda u_1(t) = f(t, \alpha(t)) - \lambda \alpha(t) \geq f(t, u_1(t)) - \lambda u_1(t),$$

and hence u_1 is a lower solution. Finally, observe that

$$u_1''(t) - \lambda u_1(t) = f(t, \alpha(t)) - \lambda \alpha(t) \geq f(t, \beta(t)) - \lambda \beta(t),$$

so using (3.4) and Lemma B.3 again, we conclude that $u_1 \leq \beta$.

Next, we apply an inductive procedure: placing u_1 in the role of α, we deduce that $u_1 \leq u_2 \leq \beta$, and so on. Thus, a sequence

$$\alpha \leq u_1 \leq u_2 \leq u_3 \leq \ldots \leq \beta$$

of lower solutions is obtained. As the sequence $\{u_n(t)\}$ is nondecreasing and bounded for all t, there exists a function $u : [0, 1] \to \mathbb{R}$ such that $u_n \to u$ pointwise; moreover, integrating (3.6) twice we obtain

$$u_{n+1}(t) = u_{n+1}'(0)t + \int_0^t \int_0^s \lambda (u_{n+1}(r) - u_n(r)) + f(r, u_n(r)) \, dr \, ds. \qquad (3.7)$$

Note that $u_n(t)$ and $u(t)$ lie between $\alpha(t)$ and $\beta(t)$ for all t; in particular, $f(\cdot, u_n)$ and $f(\cdot, u)$ are bounded functions. Thus, we may use the Lebesgue dominated convergence theorem in order to verify that the integral term converges to $\int_0^t \int_0^s f(r, u(r)) \, dr \, ds$. Taking the limit at both sides of the equality we deduce that $u_n'(0)$ converges to some constant a, and

$$u(t) = at + \int_0^t \int_0^s f(r, u(r)) \, dr \, ds.$$

This implies that u is continuous, $u''(t) = f(t, u(t))$ and, consequently, that $u \in C^2([0, 1])$ The Dirichlet condition is obviously satisfied since $u_n(0) = u_n(1) = 0$ for all n. Notice that the monotonicity implies (by Dini's theorem) that the convergence of $\{u_n\}$ is uniform, but using directly from identity (3.7) we obtain the stronger conclusion that, in fact, $u_n \to u$ for the C^2 norm. $\square$

The method of upper and lower solutions can be readily adapted to the nonhomogeneous case and to other boundary conditions (Exercise 3.2). As we shall see in next chapter, Theorem 3.2 is still true when f is only continuous.

It is interesting to observe that, unlike the elementary one-dimensional case treated at the beginning of the chapter, it is not generally true that solutions exist when α and β are in the reverse order. Consider, for instance, the problem

$$u''(t) + 4\pi^2 u(t) = \sin(2\pi t), \qquad u(0) = u(1) = 0.$$

Suppose that u is a solution, multiply the equation by $\sin(2\pi t)$, and integrate by parts twice to obtain

$$-4\pi^2 \int_0^1 u(t)\sin(2\pi t)\,dt + 4\pi^2 \int_0^1 u(t)\sin(2\pi t)\,dt = \int_0^1 \sin^2(2\pi t)\,dt,$$

a contradiction. Note, however, that $\alpha(t) = \sin(\pi t)$ and $\beta(t) = -\sin(\pi t)$ are respectively a lower and an upper solution, with $\alpha \geq \beta$.

The situation is different when the method is used for a first-order problem under periodic conditions:

$$u'(t) = f(t, u(t)), \qquad u(0) = u(1). \tag{3.8}$$

Here, if $X := \{u \in C^1([0,1]) : u(0) = u(1)\}$, then the operator $L : X \to C([0,1])$ given by $Lu := u'$ has a nontrivial kernel, the set of constant functions. Nevertheless, it becomes invertible (exercise!) if we add a term λu for any constant $\lambda \neq 0$. But the most interesting property is that the inverse of the operator $L_\lambda u := u' + \lambda u$ is nondecreasing when $\lambda > 0$ and nonincreasing when $\lambda < 0$.

Lemma 3.1. *Let $u, v \in X$ such that $u' + \lambda u \leq v' + \lambda v$. Then $u \leq v$ if $\lambda > 0$, and $u \geq v$ if $\lambda < 0$.*

Proof. Let $w = u - v$. For $\lambda > 0$, fix $t_0 \in (0, 1]$ such that $w(t_0) = \max_{t \in [0,1]} w(t)$; then $w'(t_0) \geq 0$. If $w(t_0) > 0$, then $w'(t_0) \leq -\lambda w(t_0) < 0$, a contradiction. The case $\lambda < 0$ follows in a similar way. $\square$

The preceding lemma will allow us to adapt the idea of upper and lower solutions to best suit our purposes. Assume that of class f is C^1 with respect to u and that $\alpha, \beta \in X$ verify $\alpha \leq \beta$ and

$$\alpha'(t) \leq f(t, \alpha(t)), \qquad \beta'(t) \geq f(t, \beta(t)). \tag{3.9}$$

Taking $\lambda > 0$ large enough, for each t the function $f(t, u) + \lambda u$ is nondecreasing for $u \in [\alpha(t), \beta(t)]$. Hence, if we set $N_\lambda u(t) := f(t, u(t)) + \lambda u(t)$, then the operator $L_\lambda^{-1} N_\lambda$ is nondecreasing and

$$\alpha \leq L_\lambda^{-1} N_\lambda \alpha, \qquad \beta \geq L_\lambda^{-1} N_\lambda \beta.$$

The same conclusion is also obtained if the opposite inequalities hold:

$$\alpha'(t) \geq f(t, \alpha(t)), \qquad \beta'(t) \leq f(t, \beta(t)), \tag{3.10}$$

now taking $\lambda \ll 0$. This is obviously due to the fact that the composition of two nonincreasing operators is nondecreasing.

We may summarize all the preceding comments in a single result, as follows.

Theorem 3.3. *Let $f : [0,1] \times \mathbb{R} \to \mathbb{R}$ be continuous and of class C^1 with respect to u, and assume that $\alpha \leq \beta$ are C^1 functions such that $\alpha(0) = \alpha(1)$ and $\beta(0) = \beta(1)$. If (3.9) or (3.10) holds, then problem (3.8) admits at least one solution u such that $\alpha \leq u \leq \beta$.*

Proof. As an exercise, the reader might verify the compactness assumption and give a proof of the result from the abstract setting of Theorem 3.1. For a direct proof, assume, for example, that (3.10) holds and define $u_0 := \alpha$ and u_{n+1} as the unique solution of the problem

$$u'_{n+1}(t) - \lambda u_{n+1}(t) = f(t, u_n(t)) - \lambda u_n(t), \qquad u_{n+1}(0) = u_{n+1}(1),$$

where $\lambda > 0$ satisfies (3.5).

From (3.10) it is seen that $u'_1 - \lambda u_1 \le \alpha' - \lambda \alpha$, so we deduce from Lemma 3.1 that $u_1 \ge \alpha$. This in turn implies $f(\cdot, \alpha) - \lambda \alpha \ge f(\cdot, u_1) - \lambda u_1$, and hence $u'_1(t) \ge f(t, u_1(t))$. On the other hand, the inequality $f(\cdot, \alpha) - \lambda \alpha \ge f(\cdot, \beta) - \lambda \beta$ implies that $u'_1(t) - \lambda u_1(t) \ge \beta'(t) - \lambda \beta(t)$, so $u_1 \le \beta$. Inductively, it is shown that $u_1 \le u_2 \le \ldots \le \beta$, from which it is easy to conclude as before that $\{u_n\}$ converges to a solution. An analogous proof is obtained if we start the sequence at $v_0 := \beta$. $\quad\square$

3.1.4 Unbounded Domains: A Diagonal Argument

In the preceding sections, we solved boundary value problems for first- or second-order problems. In all cases, we considered, for simplicity, only the domain $(0, 1)$ although it is clear that the results could be extended to an arbitrary bounded interval (a, b) in a straightforward manner. The reader may verify this fact by copying the previous arguments in $C([a, b])$ or by simply using the change of variables $s = t(b - a) + a$. For the same reason, in the Dirichlet problem we preferred homogeneous conditions, but nothing changes much if we consider instead nonhomogeneous ones.

However, the problem is essentially different when the domain is unbounded. In this section, we shall develop a simple method that will allow us to find solutions to some boundary problems in the half-line, in the presence of an ordered couple of a lower and an upper solution. For convenience, we shall assume that $f : [0, +\infty) \times \mathbb{R} \to \mathbb{R}$ is continuous and of class C^1 with respect to u, although the result can be generalized when f is only continuous. The key idea of the method shall be the celebrated Cantor's diagonal argument.

Let us consider again the second-order equation, now for $t > 0$, under boundary conditions on $[0, +\infty)$:

$$\begin{cases} u''(t) = f(t, u(t)), & t \in (0, +\infty), \\ u(0) = u_0, \\ u(\infty) := \lim_{t \to +\infty} u(t) = u_\infty. \end{cases} \tag{3.11}$$

Moreover, assume that $\alpha, \beta \in C^2([0, +\infty))$ are respectively a lower and an upper solution of the problem, that is,

$$\alpha''(t) \ge f(t, \alpha(t)), \qquad \beta''(t) \le f(t, \beta(t)),$$

$$\alpha(0) \le u_0 \le \beta(0), \qquad \alpha(\infty) = \beta(\infty) = u_\infty.$$

We may start by solving, for each $N \in \mathbb{N}$, a Dirichlet problem over the bounded interval $[0,N]$. Observe that the condition at 0 is clearly u_0, but we do not have a prescribed Dirichlet condition u_N for the solution at the point N. In this sense, we are free to choose the one we like most, although if we want to use the method of upper and lower solutions with α and β, then we should impose that $\alpha(N) \leq u_N \leq \beta(N)$. For example, a reasonable choice is the middle point:

$$\begin{cases} u''(t) = f(t,u(t)), & t \in (0,N), \\ u(0) = u_0, \\ u(N) = \frac{\alpha(N)+\beta(N)}{2}. \end{cases} \tag{3.12}$$

This problem has (at least) a solution v_N with $\alpha \leq v_N \leq \beta$ in $[0,N]$; however, it is not necessarily true that $v_{N+1}|_{[0,N]} = v_N$, so there is not much hope of finding a solution $v : [0,+\infty) \to \mathbb{R}$ that extends all these functions at once.

Making its appearance on the scene here is the diagonal argument. First, note that, for every M, the sequence $\{v_N\}_{N \geq M}$ is bounded; more specifically, if φ_N is the line joining the points $(0,u_0)$ and $(M,u_N(M))$, then for all $t \in [0,M]$

$$|(v_N - \varphi_N)''(t)| = |f(t,v_N(t))| \leq \max_{t \in [0,M], \alpha(t) \leq u \leq \beta(t)} |f(t,u)| := c_M.$$

Moreover, writing

$$[v_N - \varphi_N](t) = \int_0^t [v_N - \varphi_N]'(s)\,ds, \qquad [v_N - \varphi_N]'(t) = \int_0^t [v_N - \varphi_N]''(s)\,ds,$$

it is immediately seen that the sequence $\{v_N\}_{N \geq M}$ is bounded in $C^2([0,M])$. By Arzelá–Ascoli's theorem, we deduce the existence of a subsequence that converges over $[0,M]$ for the C^1 norm.

Next, we proceed as follows. Let $M = 1$, and choose a subsequence, still denoted $\{v_N\}$, that converges in $C^1([0,1])$ to some function u^1. Repeating the procedure for $M = 2,3,\ldots$, we may assume that $v_N|_{[0,M]}$ converges to some function v^M for the C^1 norm.

It is clear from the construction that $v^{M+1}|_{[0,M]} = v^M$; in particular, the function $u : [0,+\infty) \to \mathbb{R}$ given by

$$u(t) = v^M(t) \quad \text{if} \quad 0 \leq t \leq M$$

is well defined. Moreover, $u(0) = u_0$, and for each M fixed, v_N'' converges uniformly in $[0,M]$ to $f(\cdot,u)$. In consequence, for $t \leq M \leq N$ we may write

$$v_N(t) = u_0 + v_N'(0)t + \int_0^t \int_0^s v_N''(r)\,dr\,ds,$$

and letting $N \to \infty$ we obtain

$$u(t) = u_0 + u'(0)t + \int_0^t \int_0^s f(r,u)\,dr\,ds.$$

In other words, u satisfies the equation over $[0,M]$, and this holds for each M, so u is a solution of the equation in $[0,+\infty)$. Finally, $\lim_{t\to+\infty} u(t) = u_\infty$ since $u(t)$ lies between $\alpha(t)$ and $\beta(t)$ for all t. Thus we have proven the following theorem.

Theorem 3.4. *Let α and β be as before with $\alpha \le \beta$. Then problem (3.11) has at least one solution u such that $\alpha \le u \le \beta$.*

3.2 Newton Method

In the previous section, we developed an iterative method that allowed us to obtain a solution to some problems in the presence of an ordered couple of a lower and an upper solution.

As shown, the method is very elementary; for example, when applied to (3.1) the iteration only requires solving, at each step, a *linear* problem. Furthermore, it is possible to use the *Green representation formula* (see appendix, Sect. B.2.1)

$$u_{n+1}(t) = \int_0^1 G_\lambda(t,s)[f(s,u_n(s)) - \lambda u_n(s)]\, ds, \tag{3.13}$$

where G_λ is the Green function associated to the operator $L_\lambda u := u'' - \lambda u$. Note that G_λ depends only on λ and not on n; in this sense, each iteration step is reduced to compute an integral.

However, the method has the disadvantage that no error are provided. In general, such estimates can be obtained if stronger conditions on f are imposed: for example, Picard's method of successive approximations guarantees linear convergence, under the (quite strong) assumption that f is Lipschitz in u with small constant.

In this section, we shall introduce a method that, under appropriate assumptions, yields to *quadratic* convergence. In fact, the method is simply a generalization of the well-known Newton method, the same one we learned in our first calculus course. Roughly speaking, it can be summarized as follows: if you want to find the zeros of a (smooth) real function F, then start at some u_0 and define u_1 as the first coordinate of the point where the line tangent to the graph of F at $(u_0, F(u_0))$ intersects the x-axis. Then repeat the procedure, and if you are lucky enough (that is, if appropriate assumptions are satisfied), then the sequence converges to a root of F. A simple computation shows that

$$u_{n+1} = u_n - \frac{F(u_n)}{F'(u_n)},$$

and in particular, very soon one realizes that the assumptions should guarantee, in the first place, that $F'(u_n) \ne 0$ for all n.

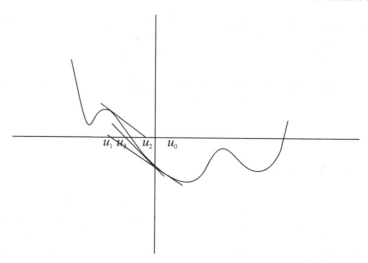

If we want to extend Newton's method to problems like (3.1), then we may search for zeros of the function $F(u) := u'' - f(\cdot, u)$ in some appropriate space. As it turns out in the elementary case of a scalar real function, some conditions shall be needed on the derivatives of F. To this end, it is convenient to recall from the previous chapter that if X, Y are normed spaces and $U \subset X$ is open, then a function $F : U \to Y$ is called C^1 if there exists a continuous application $DF : U \to L(X, Y)$ such that, for all $x \in U$,

$$\lim_{y \to x} \frac{F(y) - F(x) - DF(x)(y - x)}{\|y - x\|} = 0.$$

In particular, if the segment $[x, y] := \{ty + (1 - t)x : t \in [0, 1]\}$ is contained in U, then the following bound for the remainder is obtained.

Lemma 3.2. *Let $F : U \to Y$ be a C^1 mapping, and let $x, y \in U$ such that $[x, y] \subset U$. Then*

$$\|F(y) - F(x) - DF(x)(y - x)\| \leq \|y - x\| \int_0^1 \|DF(x + t(y - x)) - DF(x)\| \, dt.$$

In particular, if there is a constant M such that

$$\|DF(z) - DF(x)\| \leq M \|z - x\| \tag{3.14}$$

for all $z \in [x, y]$, then

$$\|F(y) - F(x) - DF(x)(y - x)\| \leq \frac{M}{2} \|y - x\|^2. \tag{3.15}$$

Proof. The case $Y = \{0\}$ is trivial, so from now on we may assume $Y \neq \{0\}$. For arbitrary $\varphi \in L(Y, \mathbb{R})$, consider the function $\Phi : [0, 1] \to \mathbb{R}$ given by

$$\Phi(t) = \varphi[F(x + t(y - x))].$$

Then $\Phi'(t) = \varphi[DF(x+t(y-x))(y-x)]$, and hence

$$\varphi(F(y) - F(x)) = \Phi(1) - \Phi(0) = \int_0^1 \varphi[DF(x+t(y-x))(y-x)]\,dt.$$

We deduce that

$$\varphi[F(y) - F(x) - DF(x)(y-x)] = \int_0^1 \varphi[DF(x+t(y-x))(y-x) - DF(x)(y-x)]\,dt.$$

Now using Lemma 2.3, we may take φ such that $\|\varphi\| = 1$ and

$$\varphi[F(y) - F(x) - DF(x)(y-x)] = \|F(y) - F(x) - DF(x)(y-x)\|.$$

Hence

$$\|F(y) - F(x) - DF(x)(y-x)\| \leq \int_0^1 \|\varphi\|\|[DF(x+t(y-x)) - DF(x)](y-x)\|\,dt,$$

and the proof follows. If, furthermore, (3.14) holds, then

$$\int_0^1 \|DF(x+t(y-x)) - DF(x)\|\,dt \leq \int_0^1 Mt\|y-x\|\,dt = \frac{M}{2}\|y-x\|,$$

and (3.15) is proven. □

Remark 3.1. 1. More generally, the result also holds if $DF(x)$ is replaced by an arbitrary continuous linear operator T.
2. In particular, (3.14) holds when F is of class C^2, with

$$M = \max_{\xi \in [x,y]} \|D^2 F\|.$$

This is due to the mean value theorem (Theorem 2.2) applied to the function DF; however, in this case the estimate (3.15) follows also directly from the Lagrange formula for Taylor remainder (see Exercise 3.10 for details).

The abstract Newton method for a C^1 mapping $F : X \to Y$ defines an iterative sequence starting at some $u_0 \in X$ from the recurrence

$$u_{n+1} = u_n - DF(u_n)^{-1}(F(u_n)). \tag{3.16}$$

As in the standard Newton method, one must be sure that $DF(u)$ is invertible for u in some appropriate region U and that the sequence does not abandon U. For the equation $u''(t) = f(t, u(t))$ under various boundary conditions, this requirement is transformed into specific conditions on the nonlinearity f. Moreover, (3.16) may be written as

$$F(u_n) = -DF(u_n)(u_{n+1} - u_n). \tag{3.17}$$

In particular, this identity shows that if the sequence $\{u_n\}$ converges, then (by continuity) its limit is a zero of F. Now substract from (3.17) the corresponding identity for u_n,

$$F(u_n) = -DF(u_n)(u_{n+1} - u_n) = F(u_n) - F(u_{n-1}) - DF(u_{n-1})(u_n - u_{n-1}),$$

and assume that $\|DF(v) - DF(u_{n-1})\| \leq M_n \|v - u_{n-1}\|$ for all $v \in [u_{n-1}, u_n]$. Then from Lemma 3.2 we obtain

$$\|F(u_n)\| = \|DF(u_n)(u_{n+1} - u_n)\| \leq \frac{M_n}{2} \|u_n - u_{n-1}\|^2.$$

On the other hand, we know from (3.16) that $\|u_{n+1} - u_n\| \leq C_n \|F(u_n)\|$, where $C_n = \|DF(u_n)^{-1}\|$, and hence

$$\|u_{n+1} - u_n\| \leq \frac{C_n M_n}{2} \|u_n - u_{n-1}\|^2.$$

Suppose for a moment that there exist constants C and M such that

$$C_n \leq C \quad \text{and} \quad M_n \leq M \quad \text{for all } n. \tag{3.18}$$

In this case, the following estimate is proven by induction:

$$\|u_{n+1} - u_n\| \leq \left(\frac{CM}{2}\right)^{2^n - 1} \|u_1 - u_0\|^{2^n} = A^{2^n - 1} \|u_1 - u_0\|,$$

where $A = \frac{CM}{2}\|u_1 - u_0\|$. Thus, the quadratic convergence of $\{u_n\}$ is guaranteed if $A < 1$ or, in other words, if $F(u_0)$ is close enough to 0. Indeed, because

$$\|u_1 - u_0\| = \|DF(u_0)^{-1}(F(u_0))\| \leq C\|F(u_0)\|,$$

condition $A < 1$ is satisfied if $\|F(u_0)\| < \frac{2}{C^2 M}$. In this case, it is clear from (3.17) that the limit of the sequence $\{u_n\}$ is a zero of F. Alternatively, observe that

$$\|F(u_n)\| \leq \frac{MC^2}{2} \|F(u_{n-1})\|^2 \leq \cdots \leq \left(\frac{MC^2}{2}\|F(u_0)\|\right)^{2^n - 1} \|F(u_0)\|,$$

and the same condition is obtained.

At this point, the reader may argue that the previous conditions are too restrictive since they require that $DF(u)$ is invertible for all u with $\|DF(u)^{-1}\|$ bounded, and that DF is a globally Lipschitz function. This happens only in very special cases, as shown by the following example.

Example 3.1. Consider problem (3.1), let

$$X = \{u \in C^2([0,1]) : u(0) = u(1) = 0\}, \qquad Y = C([0,1]),$$

and let the function $F : X \to Y$ be given by $F(u) := u'' - f(\cdot, u)$. If $f : [0,1] \times \mathbb{R} \to \mathbb{R}$ is continuous and of class C^2 with respect to u, then F is a C^2 function, with

$$DF(u)(\varphi) = \varphi'' - \frac{\partial f}{\partial u}(\cdot, u)\varphi.$$

In particular, if f is nondecreasing with respect to u, then, as we saw in previous examples, $DF(u)$ is invertible for all u; furthermore, observe that

$$-\int_0^1 \left(\varphi''(t) - \frac{\partial f}{\partial u}(t, u(t))\varphi(t) \right) \varphi(t)\, dt = \int_0^1 \varphi'(t)^2\, dt + \int_0^1 \frac{\partial f}{\partial u}(t, u(t))\varphi(t)^2\, dt.$$

As $\frac{\partial f}{\partial u} \geq 0$, by the Cauchy–Schwarz inequality we deduce

$$\|\varphi'\|_{L^2}^2 \leq \|\varphi\|_{L^2}\|DF(u)(\varphi)\|_{L^2},$$

and using the Poincaré inequality $\|\varphi\|_{L^2} \leq \frac{1}{\pi}\|\varphi'\|_{L^2}$ (see appendix in Sect. B.2.2), we obtain

$$\|\varphi'\|_{L^2} \leq \frac{1}{\pi}\|DF(u)(\varphi)\|_{L^2}.$$

Moreover, recall that $\|\varphi\|_\infty \leq \frac{1}{2}\|\varphi'\|_{L^2}$; hence, for $v \in Y$ we may set $\varphi = DF(u)^{-1}(v)$ in the previous inequality to conclude

$$\|DF(u)^{-1}(v)\|_\infty \leq \frac{1}{2\pi}\|v\|_{L^2} \leq \frac{1}{2\pi}\|v\|_\infty.$$

This shows that $\|DF(u)^{-1}\| \leq \frac{1}{2\pi}$. On the other hand, we saw in Example 1.2 that the (unique) solution u of (3.1) satisfies $\|u\|_\infty \leq R$ for some R depending only on $f(\cdot, 0)$. Thus, we may assume that DF is globally Lipschitz just by replacing f by a nondecreasing C^2 function $\tilde{f}$ such that $\tilde{f} \equiv f$ on $[-R, R]$ and $\frac{\partial^2 \tilde{f}}{\partial u^2}$ is bounded. This is seen as a direct consequence of the mean value theorem or, even more directly, using the second-order differentiation defined in Exercise 3.10.

Summarizing, in this case we may assume that C and M exist, but still, if we want to ensure that the Newton iteration converges, then we need to start close enough to the solution, but maybe we have no clue as to where possible solutions might be located. A way to avoid this difficulty shall be discussed in the following section.

Besides the specific examples, it is true that the previous conditions are quite restrictive, so it seems convenient to weaken them in some way. For instance, suppose that we know, for a certain $R > 0$, that $DF(u)$ is invertible for $u \in \overline{B_R(u_0)}$ and, moreover, that

$$C := \sup_{u \in \overline{B_R(u_0)}} \|DF(u)^{-1}\| < \infty.$$

Further, let us assume that DF is Lipschitz in $\overline{B_R(u_0)}$ with constant M. Observe that if $u_n \in \overline{B_R(u_0)}$, then u_{n+1} is defined, but unfortunately, we cannot ensure that $u_{n+1} \in \overline{B_R(u_0)}$. Thus, it is desirable to find a sufficient condition that guarantees that

$u_n \in \overline{B_R(u_0)}$ for all n, so the sequence is well defined and converges when $A < 1$. To find a sufficient condition, observe that if $u_1, \ldots, u_n \in \overline{B_R(u_0)}$, then

$$\|u_{n+1} - u_0\| \leq \sum_{j=0}^{n} \|u_{j+1} - u_j\| \leq \|u_1 - u_0\| \sum_{j=0}^{n} A^{2^j - 1} \leq \frac{1}{1-A} \|u_1 - u_0\|.$$

This means that if $\|u_1 - u_0\| < (1-A)R$, then $A < 1$ and $u_n \in \overline{B_R(u_0)}$ for all n. Taking into account the previous computations, it is enough to assume that

$$\|DF(u_0)^{-1}(F(u_0))\| \left(1 + \frac{CMR}{2}\right) < R. \tag{3.19}$$

Even if we know in advance that $DF(u)$ is invertible, condition (3.19) is hard to check since the constants C and M depend on R. The following result provides an easy-to-verify assumption.

Proposition 3.1. *Assume that $F : \overline{B_R(u_0)} \to Y$ is a C^1 function such that $DF(u_0)$ is invertible and DF is Lipschitz with constant M. Let*

$$C_0 := \|DF(u_0)^{-1}\|, \qquad \alpha_0 := \|DF(u_0)^{-1}(F(u_0))\|.$$

If

$$C_0 M \alpha_0 < \frac{1}{2} \qquad \text{and} \qquad \alpha_0 < \frac{R}{2}, \tag{3.20}$$

then the Newton iteration is well defined and converges quadratically to an element $u \in \overline{B_{2\alpha_0}(u_0)}$ such that $F(u) = 0$. Furthermore, F has no other zeros in $\overline{B_{2\alpha_0}(u_0)}$.

Proof. Note, in the first place, that u_1 is defined and

$$\|u_1 - u_0\| = \|DF(u_0)^{-1}(F(u_0))\| = \alpha_0.$$

Moreover, write

$$DF(u_1) = DF(u_0) + DF(u_1) - DF(u_0)$$
$$= DF(u_0) \left[I + DF(u_0)^{-1}(DF(u_1) - DF(u_0))\right],$$

and observe that

$$\|DF(u_0)^{-1}(DF(u_1) - DF(u_0))\| \leq C_0 M \|u_1 - u_0\| = C_0 M \alpha_0 < \frac{1}{2}.$$

From the always useful Lemma 2.4, if $T := DF(u_0)^{-1}(DF(u_1) - DF(u_0))$, then the operator $I + T$ is invertible and the norm of its inverse is less than or equal to $\frac{1}{1 - C_0 M \alpha_0}$. Consequently, $DF(u_1)$ is invertible and

$$\|DF(u_1)^{-1}\| \leq \frac{C_0}{1 - C_0 M \alpha_0}.$$

Now let

$$C_1 := \|DF(u_1)^{-1}\|, \qquad \alpha_1 := \|DF(u_1)^{-1}(F(u_1))\|, \qquad R_1 := \frac{R}{2};$$

then

$$\alpha_1 \le \|DF(u_1)^{-1}\|\|F(u_1)\| = C_1\|F(u_1) - [F(u_0) + DF(u_0)(u_1 - u_0)]\|.$$

The latter step may look weird, but it is just an old trick: simply note that the expression in square brackets is equal to 0. From Lemma 3.2 and the bound for C_1 previously obtained,

$$\alpha_1 \le C_1 \frac{M}{2}\|u_1 - u_0\|^2 \le \frac{C_0}{1 - C_0 M \alpha_0} \frac{M \alpha_0^2}{2} = \frac{C_0 M \alpha_0}{1 - C_0 M \alpha_0} \frac{\alpha_0}{2} < \frac{\alpha_0}{2} < \frac{R_1}{2}.$$

(Any questions? Just draw a graph of the function $\frac{x}{1-x}$). On the other hand,

$$C_1 M \alpha_1 \le C_1 M \frac{C_1 M \alpha_0^2}{2} \le \frac{1}{2}\left(\frac{C_0 M \alpha_0}{1 - C_0 M \alpha_0}\right)^2 < \frac{1}{2}.$$

This shows that the assumptions of the proposition are again satisfied, now for u_1 and $R_1 = \frac{R}{2}$. Inductively, it follows that u_n is well defined, and if

$$C_n := \|DF(u_n)^{-1}\|, \quad \alpha_n := \|DF(u_n)^{-1}(F(u_n))\| = \|u_{n+1} - u_n\|, \quad R_n := \frac{R_{n-1}}{2},$$

then $C_n M \alpha_n < \frac{1}{2}$ and

$$C_{n+1} \le \frac{C_n}{1 - C_n M \alpha_n}, \qquad \alpha_{n+1} \le \frac{C_n M \alpha_n}{1 - C_n M \alpha_n} \frac{\alpha_n}{2} < \frac{\alpha_n}{2} < \frac{R_{n+1}}{2}.$$

In particular,

$$\|u_{n+k} - u_n\| \le \sum_{j=0}^{k-1} \|u_{n+j+1} - u_{n+j}\| < \sum_{j=0}^{k-1} \frac{\alpha_0}{2^{n+j}} < \frac{\alpha_0}{2^{n-1}}.$$

This implies that $\{u_n\}$ is a Cauchy sequence and, hence, converges to some u, which is a zero of F. Furthermore, taking $n = 0$ in the previous formula and letting $k \to \infty$, it is seen that $u \in \overline{B_{2\alpha_0}(u_0)}$.

Now observe that the inequality $\alpha_{n+1} < \frac{\alpha_n}{2}$ yields only *linear* convergence, so there is still some extra work to be done. As before, we may write $DF(u)$ as $DF(u_0) + DF(u) - DF(u_0)$ to deduce that $DF(u)$ is invertible and, from the inverse function theorem (Exercise 2.10), $DF(u_n)^{-1} \to DF(u)^{-1}$. This implies that the sequence $\{C_n\}$ has an upper bound C, and hence

$$\alpha_{n+1} \leq \frac{C_n M \alpha_n}{1 - C_n M \alpha_n} \frac{\alpha_n}{2} \leq \frac{CM}{1 - C_n M \alpha_n} \frac{\alpha_n^2}{2} < CM \alpha_n^2,$$

so the convergence is, indeed, quadratic.

Finally, suppose that $v \in \overline{B_{2\alpha_0}(u_0)}$ is a solution. As $F(u) = F(v) = 0$, we may write

$$\|u - v\| = \|DF(u_0)^{-1}(DF(u_0)(u - v))\| \leq C_0 \|F(u) - F(v) - DF(u_0)(u - v)\|.$$

From Remark 3.1 with $T = DF(u_0)$ we obtain

$$\|F(u) - F(v) - DF(u_0)(u - v)\| \leq \|u - v\| \int_0^1 \|DF(v + t(u - v)) - DF(u_0)\| \, dt,$$

and hence, because $v + t(u - v) \in \overline{B_{2\alpha_0}(u_0)}$,

$$\|u - v\| \leq C_0 M \|u - v\| \int_0^1 \|v + t(u - v) - u_0\| \, dt \leq C_0 M \|u - v\| 2\alpha_0 < \|u - v\|,$$

so $v = u$. $\square$

The preceding results imply, in particular, that if u_0 is a zero of F and $DF(u_0)$ is invertible, then the equation $G(u) = 0$ has a solution near u_0 when G is close enough to F. Of course, the key is what we understand by "close enough": as stated, the claim seems to be just a consequence of the implicit function theorem. For example, the reader may prove the following particular case.

Exercise 3.1. Let $F : B_R(u_0) \to Y$ be a C^1 function such that $F(u_0) = 0$ and $DF(u_0)$ is invertible, and let $G = F + \varepsilon \phi$ for some fixed C^1 function ϕ. Prove that if $|\varepsilon|$ is small enough, then there exists a neighborhood V of u_0 such that G has exactly one zero $u(\varepsilon) \in V$.

The novelty in this chapter is that, under some extra assumptions, $u(\varepsilon)$ can be obtained by the Newton iteration. Indeed, assume that F is a C^1 function such that $F(u_0) = 0$ and $DF(u_0)$ is invertible. Furthermore, suppose that DF is Lipschitz in $B_R(u_0)$ with constant M. Next, take a C^1 function $\phi : B_R(u_0) \to Y$ such that $D\phi$ is also Lipschitz, with constant M_ϕ. Writing

$$D(F + \varepsilon \phi)(u_0) = DF(u_0) + \varepsilon D\phi(u_0) = DF(u_0)[I + \varepsilon DF(u_0)^{-1} D\phi(u_0)],$$

it follows that $D(F + \varepsilon \phi)(u_0)$ is invertible when $\varepsilon \|DF(u_0)^{-1} D\phi(u_0)\| < 1$. Moreover, if

$$C_\varepsilon := \|D(F + \varepsilon \phi)(u_0)^{-1}\| \quad \text{and} \quad \alpha_\varepsilon := \|D(F + \varepsilon \phi)(u_0)^{-1}(F + \varepsilon \phi)(u_0)),\|$$

then

$$C_\varepsilon \to \|DF(u_0)^{-1}\| \quad \text{and} \quad \alpha_\varepsilon \to 0$$

as $\varepsilon \to 0$. Furthermore, $D(F + \varepsilon \phi)$ is Lipschitz with constant $M_\varepsilon = M + \varepsilon M_\phi$, so we conclude that the conditions of the previous proposition are satisfied if ε is small enough.

Remark 3.2. * There are more "novelties" in this chapter concerning the implicit function theorem, specifically, that the implicit function can be obtained by the simplified Newton method (which yields linear convergence) or, under stronger assumptions, by the standard one. This is left as an exercise (see [27] for details).

Exercise 3.2. Let $F : U \times V \to Z$ be a C^1 mapping, where U and V are respectively open subsets of Banach spaces X and Y. Assume that $F(x_0, y_0) = 0$ and that $\frac{\partial F}{\partial y}(x_0, y_0)$ is invertible.

1. Prove that the iterates

$$y_{n+1}(x) = y_n(x) - \frac{\partial F}{\partial y}(x_0, y_0)^{-1}(F(x, y_n(x)))$$

   are well defined and converge to a solution of the implicit equation $F(x, y(x)) = 0$ for x in a neighborhood of x_0.
2. Assume, furthermore, that $\frac{\partial F}{\partial y}$ is Lipschitz in y in a neighborhood of (x_0, y_0), and prove that the iterates

$$y_{n+1}(x) = y_n(x) - \frac{\partial F}{\partial y}(x, y_n)^{-1}(F(x, y_n(x)))$$

   are well defined and converge quadratically to a solution of the implicit equation $F(x, y(x)) = 0$ for x in a neighborhood of x_0.

When the method is applied to problems like (3.1), some of the assumptions may be explicitly specified in terms of f. Suppose that f is continuous and of class C^1 with respect to u. Then the Newton iteration (3.17) is defined from the linear problems

$$\begin{cases} u_{n+1}''(t) = f(t, u_n(t)) + \frac{\partial f}{\partial u}(t, u_n(t))(u_{n+1}(t) - u_n(t)), \\ u_{n+1}(0) = u_{n+1}(1) = 0. \end{cases} \tag{3.21}$$

Ideally, this requires knowing in advance that $DF(u_n)$ is invertible, although sometimes it is possible to assume less. As we saw in Proposition 3.1, a simple way to ensure the success of Newton's method consists in assuming only that $DF(u_0)$ is invertible, provided that DF is Lipschitz in a neighborhood of u_0 and $F(u_0)$ is small enough. The first condition is verified, for example, if $\frac{\partial f}{\partial u}(t, u_0(t)) \geq 0$ for all t.[1]

---

[1] More generally, we might assume that $\frac{\partial f}{\partial u}(t, u_0(t))$ does not intersect the *spectrum* of the operator u'' with Dirichlet conditions, that is, $\frac{\partial f}{\partial u}(t, u_0(t)) \in \mathbb{R} \setminus \{-(n\pi)^2 : n \in \mathbb{N}\}$ for all t. (see appendix, Sect. B.2.2).

We may also assume, for simplicity, that f is of class C^2 with respect to u. Let $R > 0$, and set

$$M := \max_{(t,u) \in K} \left| \frac{\partial^2 f}{\partial u^2}(t,u) \right|,$$

where $K := \{(t,u) : t \in [0,1], |u - u_0(t)| \leq R\}$. Then DF is Lipschitz on $\overline{B_R(u_0)}$ with constant M. Moreover, we know from Example 3.1 that $\|DF(u_0)^{-1}\| \leq \frac{1}{2\pi}$ and, consequently, $\|DF(u_0)^{-1}(F(u_0))\| \leq \frac{1}{2\pi}\|F(u_0)\|$. Thus, the assumptions of Proposition 3.1 are satisfied if

$$\|F(u_0)\| < \min\left\{ \frac{\pi^2}{M}, R\pi \right\}.$$

3.2.1 Newton Continuation Method

As we saw in the preceding section, besides other conditions, Newton's method forces the iteration to start close enough to a solution. This is not always easy to verify. In this section, we shall introduce a new parameter into the problem that will allow us, in some cases, to avoid this difficulty.

Suppose we want to find a solution to the equation $F(u) = 0$, and assume we know a zero u_0 of some other function F_0. The idea of the method consists in defining a continuous *homotopy* $h : U \times [0,1] \to Y$ between F_0 and F, namely, such that $h(u,0) = F_0(u)$ and $h(u,1) = F(u)$. Starting at a zero u_λ of the function $F_\lambda := h(\cdot, \lambda)$ and making appropriate assumptions, Newton's method converges to a zero of $F_{\lambda+\varepsilon}$ for some $\varepsilon > 0$. Repeating the process, a sequence

$$0 = \lambda_0 < \lambda_1 < \lambda_2 < \dots$$

is thus obtained such that F_{λ_j} has a zero u_{λ_j} for each j. But life is not always rosy: because ε depends on λ, usually extra assumptions are needed in order to ensure that solutions can be continued up to $\lambda = 1$.

For simplicity, we shall consider only the linear homotopy

$$F_\lambda(u) = \lambda F(u) + (1 - \lambda) F_0(u),$$

where F and F_0 are C^1 mappings. Assume that u_λ is a zero of F_λ for some $\lambda \in [0,1)$ and that $DF_\lambda(u_\lambda)$ is invertible. Setting $\phi = F - F_0$, we obtain

$$DF_{\lambda+\varepsilon}(u_\lambda) = DF_\lambda(u_\lambda)\left[I + \varepsilon DF_\lambda(u_\lambda)^{-1}D\phi(u_\lambda)\right].$$

By Lemma 2.4, $DF_{\lambda+\varepsilon}(u_\lambda)$ is invertible, provided that

$$k_\varepsilon := \varepsilon\|DF_{\lambda+\varepsilon}(u_\lambda)^{-1}D\phi(u_\lambda)\| < 1.$$

In this case,

$$C_\varepsilon := \|DF_{\lambda+\varepsilon}(u_\lambda)^{-1}\| \leq \frac{\|DF_\lambda(u_\lambda)^{-1}\|}{1 - k_\varepsilon}$$

and

$$\alpha_\varepsilon := \|DF_{\lambda+\varepsilon}(u_\lambda)^{-1}(F_{\lambda+\varepsilon}(u_\lambda))\| = \|DF_{\lambda+\varepsilon}(u_\lambda)^{-1}(F_{\lambda+\varepsilon}(u_\lambda) - F_\lambda(u_\lambda))\|$$

$$= \|DF_{\lambda+\varepsilon}(u_\lambda)^{-1}(\varepsilon(F(u_\lambda) - F_0(u_\lambda)))\| \le \varepsilon C_\varepsilon \|\phi(u_\lambda)\|.$$

Now assume that F_λ and ϕ are Lipschitz on $\overline{B_R(u_\lambda)}$ with respective constants M_λ and M_ϕ; then $F_{\lambda+\varepsilon}$ is Lipschitz with constant $M_\varepsilon := M_\lambda + \varepsilon M_\phi$. From Proposition 3.1, the Newton iteration converges to a zero of $F_{\lambda+\varepsilon}$, provided that

$$\varepsilon C_\varepsilon^2 \|\phi(u_\lambda)\| \|M_\varepsilon < \frac{1}{2}, \qquad \varepsilon C_\varepsilon \|\phi(u_\lambda)\| < \frac{R}{2}.$$

As a simple application, we may consider again the Dirichlet problem for the forced pendulum equation

$$u''(t) + \sin(u(t)) = p(t), \qquad u(0) = u(1) = 0.$$

Set X as in Example 3.1, $F_\lambda u := u'' + \lambda(\sin u - p)$, and compute $DF_\lambda(u)(\varphi) = \varphi'' + \lambda \cos(u)\varphi$. As the reader may verify, $DF_\lambda(u)$ is invertible for all $\lambda \in [0,1]$ and all u; furthermore, using integration by parts and the Poincaré inequality

$$-\int_0^1 (\varphi''(t) + \lambda\cos(u(t))\varphi(t))\varphi(t)\,dt \ge \|\varphi'\|_{L^2}^2 - \|\varphi\|_{L^2}^2 \ge \|\varphi'\|_{L^2}^2\left(1 - \frac{1}{\pi^2}\right).$$

Hence, by the Cauchy–Schwarz inequality,

$$\|\varphi'\|_{L^2}^2 \le \frac{\pi^2}{\pi^2 - 1}\|DF_\lambda(u)(\varphi)\|_{L^2}\|\varphi\|_{L^2} \le \frac{\pi}{\pi^2 - 1}\|DF_\lambda(u)(\varphi)\|_{L^2}\|\varphi'\|_{L^2}.$$

Because $\|\varphi\|_\infty \le \frac{1}{2}\|\varphi'\|_{L^2}$, we deduce that $\|DF_\lambda(u)^{-1}\| \le \frac{\pi}{2(\pi^2-1)}$. Moreover,

$$[DF_\lambda(u) - DF_\lambda(v)](\varphi) = \lambda(\cos(u) - \cos(v))\varphi = -\lambda\sin(\xi)(u - v)\varphi$$

for some mean value function ξ, so DF_λ is Lipschitz with constant $L \le 1$. Finally, $\phi(u) = F_1(u) - F_0(u) = \sin(u) - p$, which implies $\|\phi(u)\| \le 1 + \|p\|_\infty$ for all u. Hence, ε may be chosen independently of λ and solutions can be continued up to the value $\lambda = 1$.

3.3 The Day When Monotone Iterations Became Quadratic

In the first two sections of this chapter, we proved the convergence of two iterative schemes: the first one, very simple, is based on the existence of an ordered couple of a lower and an upper solution and converges at a nonspecified rate; the second one yields quadratic convergence, although it requires stronger assumptions. In this section, we shall present a new method, called *quasilinearization*, that, in some sense, takes out the best of both methods. For simplicity, we shall restrict ourselves to the particular case (3.1); a general version of the method can be obtained from this outline.

Assume that $f : [0,1] \times \mathbb{R} \to \mathbb{R}$ is continuous and of class C^2 with respect to u, and let $\alpha \leq \beta$ respectively be a lower and an upper solution. In contrast with the previous methods, we shall solve, at each step, a *quasilinear* problem. However, we shall prove a posteriori that the iteration coincides with that given by in Newton's method, which is defined by a *linear* problem. With this aim, let P be the truncation function defined by

$$P(t,u) = \begin{cases} u & \text{if } \alpha(t) \leq u \leq \beta(t), \\ \alpha(t) & \text{if } u < \alpha(t), \\ \beta(t) & \text{if } u > \beta(t), \end{cases}$$

and set $u_0 := \alpha$. Moreover, let $\mathscr{C} := \{(t,u) : t \in [0,1], \alpha(t) \leq u \leq \beta(t)\}$, and fix $\lambda \geq 0$ such that $\lambda \geq \frac{\partial f}{\partial u}(t,u)$ for all $(t,u) \in \mathscr{C}$. Next, consider the quasilinear problem

$$\begin{cases} u''(t) - \lambda u(t) = f(t,\alpha(t)) - \lambda P(t,u(t)) + \frac{\partial f}{\partial u}(t,\alpha(t))[P(t,u(t)) - \alpha(t)], \\ u(0) = u(1) = 0. \end{cases}$$

$$(3.22)$$

Because the right-hand-side term is bounded, we know from Chap. 1 that (3.22) has a solution u, which is not necessarily unique. Furthermore,

$$u''(t) - \lambda u(t) = f(t,\alpha(t)) - \lambda\alpha(t) + \left(\frac{\partial f}{\partial u}(t,\alpha(t)) - \lambda\right)[P(t,u(t)) - \alpha(t)]$$

$$\leq f(t,\alpha(t)) - \lambda\alpha(t) \leq \alpha''(t) - \lambda\alpha(t),$$

so we deduce from the maximum principle that $u \geq \alpha$.

Next, assume that $\frac{\partial^2 f}{\partial u^2}(t,u) \leq 0$ for $(t,u) \in \mathscr{C}$, and consider the Taylor expansion

$$f(t,P(t,u(t))) = f(t,\alpha(t)) + \frac{\partial f}{\partial u}(t,\alpha(t))(P(t,u(t)) - \alpha(t)) + R(P(t,u(t))),$$

$$(3.23)$$

where, from the previous concavity assumption, $R(P(t,u(t))) \leq 0$. Using also the fact that $f(t,u) - \lambda u$ is nonincreasing we obtain

$$u''(t) - \lambda u(t) = f(t,P(t,u(t))) - \lambda P(t,u(t)) - R(P(t,u(t)))$$

$$\geq f(t,\beta(t)) - \lambda\beta(t) \geq \beta''(t) - \lambda\beta(t),$$

and we conclude that $u \leq \beta$.

Once we know that $\alpha \le u \le \beta$, then (3.22) simply reads

$$u''(t) = f(t, \alpha(t)) + \frac{\partial f}{\partial u}(t, \alpha(t))(u(t) - \alpha(t)),$$

and moreover, by (3.23), the latter term is equal to

$$f(t, u(t)) - R(u(t)) \ge f(t, u(t)).$$

Thus, u is a lower solution and the procedure may be repeated. In this way, we obtain a sequence

$$\alpha = u_0 \le u_1 \le u_2 \le \cdots \le \beta$$

of lower solutions such that

$$\begin{cases} u''_{n+1}(t) = f(t, u_n(t)) + \frac{\partial f}{\partial u}(t, u_n(t))(u_{n+1}(t) - u_n(t)), \\ u_{n+1}(0) = u_{n+1}(1) = 0. \end{cases} \qquad (3.24)$$

Note that the sequence is not uniquely defined, but, nevertheless, (3.24) coincides with the iteration defined by Newton's method. The exercise of proving that u_n converges to a C^2 solution of the problem is left to the reader. Observe that the truncation function is redefined at each step, so it depends on n. Also, it is worth noting that the sequence cannot be defined directly from (3.24) since we do not know in advance that the problem is solvable.

As mentioned, a condition that would ensure the (unique) solvability of (3.24) is

$$\frac{\partial f}{\partial u}(t, u) \ge c > -\pi^2 \qquad (3.25)$$

for all (t, u). Fortunately, this condition was not used at all in the previous construction since, in that case, the whole quasilinearization adventure would have become a waste of time. Nevertheless, if we restrict (3.25) just to those elements $(t, u) \in \mathscr{C}$, then the iteration is not reduced a priori to the standard Newton case, and yet the previous sequence still converges quadratically.

Indeed, let $\varepsilon_n := u - u_n$ be the nth error term; then $\varepsilon_n \ge 0$ and, moreover,

$$\varepsilon_n''(t) = f(t, u(t)) - f(t, u_{n-1}(t)) - \frac{\partial f}{\partial u}(t, u_{n-1}(t))(u_n(t) - u_{n-1}(t))$$

$$\varepsilon_n(0) = \varepsilon_n(1) = 0.$$

Hence

$$\varepsilon_n''(t) - \frac{\partial f}{\partial u}(t, u_{n-1}(t))\varepsilon_n(t) = f(t, u(t)) - f(t, u_{n-1}(t)) - \frac{\partial f}{\partial u}(t, u_{n-1}(t))\varepsilon_{n-1}(t)$$

$$= \frac{1}{2}\frac{\partial^2 f}{\partial u^2}(t, \xi_n(t))\varepsilon_{n-1}(t)^2$$

for some mean value $\xi_n(t)$.

Next, define ϕ_n as the unique solution of the problem

$$\phi_n''(t) - c\phi_n(t) = \varepsilon_n''(t) - \frac{\partial f}{\partial u}(t, u_{n-1}(t))\varepsilon_n(t)$$

$$\phi_n(0) = \phi_n(1) = 0,$$

with c as in (3.25); then

$$\varepsilon_n'' - c\varepsilon_n = \varepsilon_n'' - \frac{\partial f}{\partial u}(\cdot, u_{n-1})\varepsilon_n + \left(\frac{\partial f}{\partial u}(\cdot, u_{n-1}) - c\right)\varepsilon_n \geq \phi_n'' - c\phi_n,$$

and from the maximum principle (see appendix, Remark B.1), $\varepsilon_n \leq \phi_n$. Moreover,

$$\|\phi_n\|_{L^2}\|\phi_n'' - c\phi_n\|_{L^2} \geq -\int_0^1 \phi_n(t)[\phi_n''(t) - c\phi_n(t)]\,dt = \int_0^1 [\phi_n'(t)^2 + c\phi_n(t)^2]\,dt,$$

so from the Poincaré inequality we deduce that

$$\left(1 + \frac{c}{\pi^2}\right)\|\phi_n'\|_{L^2}^2 \leq \|\phi_n\|_{L^2}\|\phi_n'' - c\phi_n\|_{L^2} \leq \frac{1}{\pi}\|\phi_n'\|_{L^2}\|\phi_n'' - c\phi_n\|_{L^2},$$

and hence $\|\phi_n'\|_{L^2} \leq \frac{\pi}{\pi^2+c}\|\phi_n'' - c\phi_n\|_{L^2} \leq \frac{\pi}{\pi^2+c}\|\phi_n'' - c\phi_n\|_\infty$. We conclude that

$$\varepsilon_n \leq \phi_n \leq \frac{1}{2}\|\phi_n'\|_{L^2} \leq \frac{M\pi}{2(\pi^2 + c)}\|\varepsilon_{n-1}\|_\infty^2,$$

where $M := \max_{(t,u)\in\mathscr{C}} \frac{\partial^2 f}{\partial u^2}(t, u)$. Summarizing, we have proven the following theorem.

Theorem 3.5. *Let $f : [0,1] \times \mathbb{R} \to \mathbb{R}$ be continuous and of class C^2 with respect to u, and let $\alpha \leq \beta$ be a lower and an upper solution of (3.1). Further, assume that $\frac{\partial^2 f}{\partial u^2}(t, u) \leq 0$ for all $(t, u) \in \mathscr{C}$. Then the sequence defined by (3.24) converges to a solution. If, moreover, (3.25) holds for all $(t, u) \in \mathscr{C}$, then the convergence is quadratic.*

Appendix

Some bibliographical notes. The method of upper solutions is widely used in the literature. Here, we have just presented a brief account of those specific aspects concerning monotone iterations; a more general overview of the method and its connections with the fixed point theorems shall be introduced in the next chapter. A good introduction to cones and orders in Banach spaces can be found, for example, in [27, 109] and in [48], where the monotone iterative technique is used for problems with discontinuous nonlinearities. The diagonalization argument is one of the most celebrated methods for boundary value problems in infinite intervals.

For second-order problems, it was used, for example, in [14] and [40]. More on infinite interval problems is said in [1]. The fundamentals of Newton's method, as introduced in this chapter, are revealed in [27] and [109], among many other works. The quasilinearization method has been applied to many different nonlinear problems in the presence of an ordered couple of a lower and an upper solution. It was first developed by Bellman and Kalaba [16] and generalized by Lakshmikantham [63, 64]. (See also the monograph [65]. For history and other issues, see [62].)

Problems

3.1. Prove that the problem

$$\begin{cases} u'(t) = u(t)^4 + \cos(t) - 1 \\ u(0) = u(2\pi) \end{cases}$$

has at least one solution u such that $\sin(t) \le u(t) \le \sin(t) + 2$ for all $t \in [0, 2\pi]$.

3.2. Generalize the method of upper and lower solutions for the scalar equation $u''(t) + a(t)u'(t) = f(t, u(t))$ with:

1. Nonhomogeneous Dirichlet conditions,
2. Neumann conditions,
3. Periodic conditions, assuming that α and β satisfy

$$\alpha(0) - \alpha(1) = \beta(0) - \beta(1) = 0, \qquad \alpha'(0) - \alpha'(1) \ge 0 \ge \beta'(0) - \beta'(1).$$

3.3. Prove that the problem

$$u''(t) = p(t) + u(t)^5 + g(t, u(t)),$$

with $p : [0, T] \to \mathbb{R}$ continuous and $g : [0, T] \times \mathbb{R} \to \mathbb{R}$ bounded continuous and C^1 in u, has at least one T-periodic solution.

3.4. Prove that the periodic problem for the forced pendulum equation with friction

$$u''(t) + au'(t) + \sin u(t) = p(t), \qquad u(0) = u(1), u'(0) = u'(1)$$

admits at least one solution for any $p \in C([0, 1])$ such that $\|p\|_\infty \le 1$.

3.5. (Adapted from [62])

1. Let X be a metric space, and let $T : X \to X$ be such that $\overline{T(X)}$ is compact and

$$d(T(x), T(y)) \le G(d(x, y)) \qquad \text{if } d(x, y) \le R,$$

where $G : [0, R] \to [0, R)$ is continuous and satisfies the following conditions:

- G is nondecreasing,
- $G(r) = r \iff r = 0$.

Prove that if $d(T(u_0), u_0) < R$ for some $u_0 \in X$, then the successive approximations scheme defined by $u_{n+1} = T(u_n)$ converges to a fixed point of T. Is it unique?

2. Reformulate Exercise 2.6 in terms of 1.

3.6. Prove that the problem

$$\begin{cases} u''(t) = u(t)^3 + u(t) \\ u(0) = u_0 \\ \lim_{t \to +\infty} u(t) = 0 \end{cases}$$

has at least one solution. Could it have more than one?

3.7. Let $f : [0, +\infty) \times \mathbb{R} \to \mathbb{R}$ be continuous and of class C^1 with respect to u such that $f(t, u)\mathrm{sgn}(u) \geq 0$ for $|u| \gg 0$. Assume, moreover, that for every $M > 0$ there exists $\varphi_M \in L^1(0, +\infty)$ such that $|f(t, u)| \leq \varphi_M(t)$ for $|u| \leq M$. Prove that the problem

$$\begin{cases} u''(t) = f(t, u(t)) \\ u'(0) = u_0 \\ \lim_{t \to +\infty} u'(t) = 0 \end{cases}$$

admits at least one solution.

3.8. Write explicit Newton and quasilinearization iterative schemes for the forced pendulum equation under Dirichlet and periodic conditions.

3.9. Give a direct proof of the existence-uniqueness result in Proposition 3.1 for the particular case $X = Y = \mathbb{R}$. More precisely, let $f : [u_0 - R, u_0 + R] \to \mathbb{R}$ be continuously differentiable such that f' is Lipschitz with constant M and

$$|f(u_0)| < \min\left\{ \frac{f'(u_0)^2}{2M}, R|f'(u_0)| \right\}.$$

Then f has exactly one zero in $\left[u_0 - 2\left|\frac{f(u_0)}{f'(u_0)}\right|, u_0 + 2\left|\frac{f(u_0)}{f'(u_0)}\right| \right]$. Propose an example to convince yourself that the assumptions in Proposition 3.1 are cannot be relaxed.

3.10. Let $F : U \subset X \to Y$ be differentiable, and assume that $DF : U \to L(X, Y)$ is also differentiable. For each $x \in U$, the operator $D(DF)(x) \in L(X, L(X, Y))$ shall be denoted by $D^2F(x)$.

1. Prove that

$$D^2F(x_0)(u)(v) = D^2F(x_0)(v)(u)$$

for all $u, v \in X$. Deduce that D^2F may be interpreted as a function from U to the space $L^2(X, Y)$ of continuous bilinear symmetric forms from $X \times X$ to Y equipped with the usual norm

$$\|B\| := \sup_{\|x\|,\|y\|\leq 1} \|B(x,y)\|.$$

If D^2F is continuous, we shall say that F is a C^2 function.

2. Assume that F is a C^2 mapping, and give a direct proof of (3.15) with $M = \max_{\xi\in[x,y]} \|D^2F\|$ using the Lagrange remainder formula for real functions. An alternative proof can be obtained using the ideas of Exercise 2.7.

3. Let $f : [0,1] \times \mathbb{R} \to \mathbb{R}$ be continuous and of class C^2 with respect to u. Prove that the function $F : C^2([0,1]) \to C([0,1])$ given by $Fu := u'' - f(\cdot,u)$ is C^2, with

$$D^2F(u)(v,w) = -\frac{\partial^2 f}{\partial u^2}(\cdot,u)vw$$

for all $v,w \in C^2([0,1])$.

3.11. Show that the Newton continuation method can be successfully applied to problem (3.1) with f smooth enough and nondecreasing with respect to u.

3.12. Consider the fourth-order problem

$$u^{(4)}(t) + g(t,u(t),u''(t)) = 0, \tag{3.26}$$

with g continuous, T-periodic in t, and of class C^2 with respect to the variables u and u''. Assume there exist smooth T-periodic functions α and β such that

$$\alpha^{(4)} + g(\cdot,\alpha,\alpha'') \leq 0,$$

$$\beta^{(4)} + g(\cdot,\beta,\beta'') \geq 0,$$

and

$$\alpha'' - K\alpha \geq \beta'' - K\beta$$

for some constant $K > 0$, and moreover,

$$\frac{\partial g}{\partial u}(t,u,v) + K\frac{\partial g}{\partial u''}(t,u,v) + K^2 \leq 0$$

for $(t,u,v) \in \mathscr{C}$, where $\mathscr{C}$ is the set of all those $(t,u,v) \in [0,T] \times \mathbb{R}^2$ such that

$$\alpha(t) \leq u \leq \beta(t),$$

$$\alpha''(t) - K\alpha(t) \geq v - Ku \geq \beta''(t) - K\beta(t).$$

1. Prove that (3.26) has at least one T-periodic solution u with $\alpha \leq u \leq \beta$.

   Hint: prove the following maximum principle. Let $\lambda,\mu > 0$ satisfy $\lambda^2 \geq 4\mu$, and let $K^{\pm} = \frac{\lambda\pm\sqrt{\lambda^2-4\mu}}{2}$. If u is T-periodic and satisfies

$$u^{(4)} - \lambda u'' + \mu u \geq 0,$$

then

$$u'' - K^{\pm}u \leq 0.$$

In particular, $u \geq 0$.

2. Develop a quasilinearization method for the problem

$$u^{(4)}(t) - 2u''(t) + u(t)^5 = p(t)$$

under periodic conditions assuming that $\|p\|_{\infty}$ is small enough.

Hint: try with constants $\alpha < 0 < \beta$.

3.13. * (Antimaximum principle)

1. Compute the Green function (see appendix, Sect. B.2.1) G_{λ} associated to the operator $Lu := u'' + \lambda u$ under periodic conditions for $0 < \lambda < 4\pi^2$, and prove that if $\lambda < \frac{\pi^2}{4}$, then $G_{\lambda} \geq 0$.
2. Deduce the following *antimaximum principle*. If u is 1-periodic and $u'' + \lambda u \geq 0$ for some $0 < \lambda < \frac{\pi^2}{4}$, then $u \geq 0$.
3. Let $f : [0,1] \times \mathbb{R} \to \mathbb{R}$ be continuous, 1-periodic in t, and of class C^1 with respect to u, and let $\alpha \geq \beta$ be 1-periodic such that $\alpha''(t) \geq f(t,\alpha(t))$ and $\beta''(t) \leq f(t,\beta(t))$ for all t. If $\frac{\partial f}{\partial u}(t,u) > -\frac{\pi^2}{4}$ for all $t \in [0,1]$ and $\alpha(t) \geq u \geq \beta(t)$, then the problem $u''(t) = f(t,u(t))$ admits at least one 1-periodic solution u with $\beta \leq u \leq \alpha$.
4. Let p be continuous and 1-periodic such that $-1 \leq p \leq 1$. Prove that the forced pendulum equation $u''(t) + \sin u(t) = p(t)$ admits a 1-periodic solution u with $-\frac{\pi}{2} \leq u \leq \frac{\pi}{2}$. Is this the same solution obtained in Exercise 3.4?

Chapter 4
The Schauder Theorem and Applications

4.1 Brouwer's Theorem Revisited

In this chapter, we shall introduce a suitable extension of the Brouwer fixed point theorem, with applications to many different boundary value problems. But first we need to establish Brouwer's theorem in the general n-dimensional case. Henceforth, $\overline{B_1(0)}$ will denote the closed unit ball of $\mathbb{R}^n$.

Theorem 4.1. *(Brouwer) Any continuous mapping* $f : \overline{B_1(0)} \to \overline{B_1(0)}$ *has at least one fixed point.*

This result can be regarded as a fundamental fact in mathematics since it proves to be equivalent to the axiom of completeness of real numbers. As seen in Chap. 1, it might be thought of as a consequence of the following no-retraction theorem.

Theorem 4.2. *There are no continuous mappings* $r : \overline{B_1(0)} \to \partial B_1(0)$ *such that* $r|_{\partial B_1(0)} = id$.

Indeed, if $f : \overline{B_1(0)} \to \overline{B_1(0)}$ is continuous and has no fixed points, then a retraction may be constructed by taking, for each $x \in \overline{B_1(0)}$, the half-line that starts at $f(x)$ and passes through x. The value $r(x)$ is thus defined as the point where this half-line intersects S^{n-1}. It is clear that r is continuous and $r(x) = x$ for $x \in S^{n-1}$. In fact, many proofs of Brouwer's result rely just on the fact that there are no *smooth* retractions from $\overline{B_1(0)}$ to S^{n-1}; indeed, this implies Brouwer's theorem for smooth mappings and the general case follows easily from the Stone–Weierstrass approximation theorem in the following way. Let $f : \overline{B_1(0)} \to \overline{B_1(0)}$ be continuous, and set $q_n : \overline{B_1(0)} \to \mathbb{R}^n$ polynomials such that $\|q_n - f\|_\infty \leq \frac{1}{n}$. It is clear that $\|q_n\|_\infty \leq \frac{n+1}{n}$; thus, we may define $p_n := \frac{n}{n+1} q_n$. It follows that $p_n : \overline{B_1(0)} \to \overline{B_1(0)}$, and $p_n \to f$ uniformly. Moreover, each p_n has a fixed point x_n, and taking a subsequence we may assume that $x_n \to x \in \overline{B_1(0)}$. Thus, $f(x_n) \to f(x)$ and $f(x_n) - x_n = f(x_n) - p_n(x_n) \to 0$, that is, $f(x) - x = 0$.

But it is easy to see, as in Chap. 1, that the preceding theorems are equivalent to one another and to the following three results.

P. Amster, *Topological Methods in the Study of Boundary Value Problems*, Universitext, DOI 10.1007/978-1-4614-8893-4_4, © Springer Science+Business Media New York 2014

Theorem 4.3. *Assume that* $\phi : \overline{B}_1(0) \to \mathbb{R}^n$ *is continuous such that*

$$\phi(x) \cdot x \geq 0 \quad \text{for all } x \in \partial B_1(0).$$

Then ϕ *has a zero in* $B_1(0)$.

Theorem 4.4. *(Poincaré–Miranda) Let* $f = (f_1, \ldots, f_n) : [-1, 1]^n \to \mathbb{R}^n$ *be continuous such that*

$$f_i(x^-) \leq 0 \leq f_i(x^+)$$

for $i = 1, \ldots, n$ *and all* $x^- \in F_i^-$ *and* $x^+ \in F_i^+$, *where* $F_i^{\pm}$ *denote the faces*

$$F_i^- := \{x \in [-1, 1]^n : x_i = -1\}, \qquad F_i^+ := \{x \in [-1, 1]^n : x_i = 1\}.$$

Then there exists $x \in [-1, 1]^n$ *such that* $f(x) = 0$.

Theorem 4.5. *If* $f : \mathbb{R}^n \to \mathbb{R}^n$ *is continuous and there exists a constant* C *such that* $|f(x) - x| \leq C$ *for all* x, *then* f *has at least one zero.*

When $n = 1$, any of the previous results follows immediately from Bolzano's theorem; in contrast with that, a significant amount of work is needed for a proof in the general case. In this context, the case $n = 2$ is very privileged and, for this reason, deserved special treatment in Chap. 1: on the one hand, Brouwer's theorem is not trivial as in the one-dimensional case; on the other hand, when $n = 2$, it is possible to perform some simple proofs that fail for higher dimensions.

Nowadays, many different proofs are known for the general case, some of them quite immediate when more powerful topological techniques are employed. For instance, as we shall see in the next chapter, the theorem can be regarded just as a simple exercise in the context of topological degree theory; if one uses, instead, homology theory, then it is readily seen that there cannot be continuous retractions from the closed unit ball of $\mathbb{R}^n$ onto the unit sphere S^{n-1} since $H_{n-1}(S^{n-1}) = \mathbb{Z}$ and $H_{n-1}(\overline{B_1(0)}) = 0$.

Nonetheless, some nontechnical proofs of Brouwer's theorem in arbitrary dimension are available. We shall present just one of them in the next section, although it is worth mentioning some others that might be of interest for the reader. For example, the proof in [53] is intuitively based on the trivial fact that, if r is a C^1 retraction from $[-1, 1] \to S^0 = \{-1, 1\}$, then $0 = \int_{-1}^{1} r'(t) \, dt = r(1) - r(-1) = 2 \neq 0$. This naïve observation serves as a starting point for the general case: the author proves that, if $r : \overline{B_1(0)} \to S^{n-1}$ is a C^1 retraction, then

$$0 = \int_{B_1(0)} det(Dr(x)) \, dx \neq 0.$$

The latter assertion, of course, is nontrivial and uses an algebraic lemma and the Stokes theorem; the argument can be seen as a generalization of that given in Sect. 1.2.2 for the case $n = 2$ by Green's theorem. There is also a simple and elegant combinatorial proof based on the well-known Sperner lemma (e.g., [50]).

A straightforward generalization of Brouwer's theorem is obtained for any set K homeomorphic to the closed unit ball of $\mathbb{R}^n$: indeed, if $f : K \to K$ is continuous, then it suffices to consider a homeomorphism $h : \overline{B} \to K$, so the continuous mapping $h^{-1} \circ f \circ h : \overline{B} \to \overline{B}$ has a fixed point x. Then $y = h(x)$ is a fixed point of f. In particular, the result holds when K is a compact convex subset of a finite-dimensional normed space.

There are many applications of Brouwer's theorem or its equivalences to boundary value problems. For example, the examples and exercises proposed in Chap. 1 that used the two-dimensional version may now be generalized (Exercise 4.1). Another standard application is the following one.

Example 4.1. (Difference equations) Consider the second-order *difference equation*

$$\begin{cases} \Delta^2 y(t) = f(t, y(t)) & t = 0, \ldots, n-2, \\ y(0) = y(n) = 0, \end{cases}$$

where the difference operator Δ is defined by $\Delta y(t) = y(t+1) - y(t)$ and the mapping $f : \{0, \ldots, n-2\} \times \mathbb{R} \to \mathbb{R}$ is continuous, that is, $f(t, \cdot) : \mathbb{R} \to \mathbb{R}$ is continuous for $t = 0, \ldots, n-2$. Identifying the functions $z : \{0, \ldots, n\} \to \mathbb{R}$ with $\mathbb{R}^{n+1}$, we may regard the previous problem as a fixed point equation in $\mathbb{R}^{n+1}$: for fixed z, solve the problem

$$\begin{cases} \Delta^2 y(t) = f(t, z(t)), & t = 0, \ldots, n-2, \\ y(0) = y(n) = 0, \end{cases}$$

and define a function $F : \mathbb{R}^{n+1} \to \mathbb{R}^{n+1}$ given by $Fz := y$. It is easy to prove that F is well defined and continuous, and its fixed points are obviously solutions of the problem. As an exercise, the reader may verify that if f is nondecreasing or sublinear in its second variable, then the problem has at least one solution in $\overline{B_R(0)}$ for some R large enough.

4.1.1 Milnor–Rogers Proof

In this section, we shall present an elementary analytic proof of the Brouwer fixed point theorem, first given by Milnor and then simplified by Rogers [88, 101]. As mentioned, it suffices to prove that there are no C^1 retractions from the closure of the unit ball onto the sphere S^{n-1}.

By contradiction, suppose that $r : \overline{B_1(0)} \to S^{n-1}$ is a C^1 mapping such that $r(x) = x$ over S^{n-1}, and define, for $\lambda \in [0, 1]$, the function

$$f_\lambda(x) = (1 - \lambda)x + \lambda r(x) = x + \lambda(r(x) - x).$$

Observe that $|f_\lambda(x)| \le (1 - \lambda)|x| + \lambda|r(x)| \le 1$ for all x and $f_\lambda(x) = x$ for $x \in S^{n-1}$.

Moreover, $g(x) := r(x) - x$ is a C^1 function, so there exists a constant c such that $|g(x) - g(y)| \le c|x - y|$ for $x, y \in \overline{B_1(0)}$. If $f_\lambda(x) = f_\lambda(y)$, then

$$|x - y| = \lambda |g(x) - g(y)| \leq \lambda c |x - y|.$$

Thus, f_λ is injective for $\lambda < \frac{1}{c}$.

We claim that, if λ is small enough, then f_λ is also surjective. Indeed, let us first observe that, by the always useful Lemma 2.4, the differential $Df_\lambda = I + \lambda Dg$ is invertible for all x, provided that $\lambda \|Dg\| < 1$. Using the inverse function theorem, we deduce that $f_\lambda(B_1(0))$ is open. Suppose $f_\lambda(B_1(0)) \neq B_1(0)$, and take $z \in B_1(0) \backslash f_\lambda(B_1(0))$. The set

$$A := \{t \in [0,1] : f_\lambda(0) + t(z - f_\lambda(0)) \in f_\lambda(B_1(0))\}$$

is nonempty, so it has a supremum t_0. Let $t_k \in A$ such that $t_k \to t_0$, and choose $x_k \in B_1(0)$ such that $f_\lambda(x_k) = f_\lambda(0) + t_k(z - f_\lambda(0))$. Passing to a subsequence, we may assume that x_k converges to some $x \in \overline{B_1(0)}$.

Note that $f_\lambda(x) = f_\lambda(0) + t_0(z - f_\lambda(0))$ lies in the segment $[f_\lambda(0), z] \subset B_1(0)$; thus, because $f_\lambda|_{\partial B_1(0)} = Id$, we conclude that $x \in B_1(0)$. Then $f_\lambda(x) \in f_\lambda(B_1(0))$; in particular, $f_\lambda(x) \neq z$, and hence $t_0 < 1$. Moreover, a neighborhood of $f_\lambda(x)$ is contained in $f_\lambda(B_1(0))$, which contradicts the fact that t_0 is the supremum.

Next, define the mapping

$$\phi(\lambda) := \int_B \det(Df_\lambda(x)) \, dx = \int_B \det(I + \lambda Dg(x)) \, dx.$$

Since f_λ is a diffeomorphism for small λ, by the change of variables formula it follows that, near the origin, ϕ is constantly equal (up to a sign) to the volume of $B_1(0)$. On the other hand, it is clear that ϕ is a polynomial: thus, it is constant over $[0,1]$ and, in particular, $\phi(1) \neq 0$.

Also, $f_1 = r$, so from the equality $r(x) \cdot r(x) = 1$ we deduce, for $j = 1, \ldots, n$, that

$$\partial_j r(x) \cdot r(x) = 0.$$

This implies that $r(x) \in \ker(Dr(x))$, and hence $\det(Dr(x)) = 0$ for all $x \in B_1(0)$. We conclude that $\phi(1) = 0$, a contradiction.

4.2 Schauder's Theorem

4.2.1 An Example by Kakutani

In view of the multiple applications of the Brouwer fixed point theorem, it would be desirable to have an appropriate generalization for arbitrary Banach spaces. In this section, we shall show that the theorem cannot be extended to the infinite-dimensional case just as it stands. Indeed, an example due to Kakutani [52] indicates that an extra condition for f is required. Here, a slightly simpler form of the example is introduced; the original version is given in Exercise 4.6.

Consider the space

$$l^2 := \left\{ x := (x_n)_{n \in \mathbb{N}} : x_n \in \mathbb{R}, \sum_{n \in \mathbb{N}} x_n^2 < \infty \right\}$$

equipped with the standard norm

$$\|x\| := \sqrt{\sum_{n \in \mathbb{N}} x_n^2}$$

and the continuous function $f : \overline{B_1(0)} \subset l^2 \to l^2$ given by

$$f(x) = \left(\sqrt{1 - \|x\|^2}, x_1, x_2, \ldots \right).$$

For $x \in \overline{B_1(0)}$, it is seen that $\|f(x)\|^2 = 1 - \|x\|^2 + \|x\|^2 = 1$; thus, $f(\overline{B_1(0)})$ is contained in $\overline{B_1(0)}$ (more precisely, in its boundary). However, f has no fixed points: suppose that $f(x) = x$; then $x_{n+1} = x_n$ for all n and, from the fact that $x_n \to 0$, we deduce that $x = 0$. But $f(0) = (1, 0, 0, \ldots) \neq 0$, a contradiction.

Observe that a similar result is obtained in an arbitrary separable Hilbert space since it is isomorphic to l^2. In this sense, the main idea of the more general example given in Exercise 4.6 is essentially the same.

4.2.2 And Yet It Doesn't Move

Let us have a second look at Kakutani's example: where is the trick? It is quite clear that the secret ingredient is hidden in the shift operator, which allows us to "move" all the coordinates of an element x to the right; this is obviously not possible in a finite-dimensional space. But there are less restrictive ways of avoiding this effect; for instance, note that the shift operator is isometrically isomorphic with its image. In particular, it cannot be *compact* since, among other things, compact sets cannot contain infinite-dimensional balls.

Theorem 4.6. *(Schauder) Let E be a normed space, and let C be a convex, bounded, and closed subset of E. If $T : C \to C$ is continuous with $\overline{T(C)}$ compact, then T has at least one fixed point.*

Proof. Let $k \in \mathbb{N}$. As $\overline{T(C)}$ is compact, there exist $n = n(k) \in \mathbb{N}$ and elements $x_1, x_2, \ldots, x_n \in \overline{T(C)} \subset C$ such that

$$\overline{T(C)} \subset \bigcup_{j=1}^{n} B_{1/k}(x_j).$$

Now let $C_k \subset C$ be the convex hull of the set $\{x_1, x_2, \ldots, x_n\}$, and define the mapping $J_k : \overline{T(C)} \to C_k$ given by

$$J_k(y) = \sum_{j=1}^{n} \frac{dist(y, \overline{T(C)} - B_j)}{\sum_{i=1}^{n} dist(y, \overline{T(C)} - B_i)} x_j,$$

where $B_j = B_{1/k}(x_j)$. Note that if $y \in \overline{T(C)}$, then $y \in B_j$ for some j, and hence $dist(y, \overline{T(C)} - B_j) > 0$. Thus, J_k is well defined and clearly continuous.

On the other hand, if $y \notin B_j$, then $dist(y, \overline{T(C)} - B_j) = 0$; thus, for every j we deduce

$$dist(y, \overline{T(C)} - B_j) \|y - x_j\| \leq \frac{1}{k} dist(y, \overline{T(C)} - B_j).$$

Hence, $\|J_k(y) - y\| \leq \frac{1}{k}$ for all $y \in \overline{T(C)}$ and, in particular, $\|J_k T(x) - Tx\| \leq \frac{1}{k}$ for all $x \in C$.

From Brouwer's theorem we deduce that $J_k \circ T|_{C_k} : C_k \to C_k$ has a fixed point z_k and, by compactness, $\{Tz_k\}$ has a subsequence $\{Tz_{k_j}\}$ that converges to some $z \in \overline{T(C)} \subset C$. Moreover,

$$\|z_{k_j} - Tz_{k_j}\| = \|J_{k_j} T(z_{k_j}) - Tz_{k_j}\| \leq \frac{1}{k_j} \to 0.$$

Hence $z_{k_j} \to z$, and by continuity, $Tz = \lim_{j \to \infty} Tz_{k_j} = z$. □

Remark 4.1. It is worth noting, in the previous proof, that the convexity assumption was not used for the construction of the operator $J_k \circ T$. A straightforward extension of the argument shows that any continuous operator $T : C \to F$ such that $\overline{T(C)}$ is compact, $C \subset E$ bounded and closed, and F an arbitrary normed space can be approximated by operators of finite range. In more precise terms: for each $\varepsilon > 0$ there exists $T_\varepsilon : C \to F$ such that $\|T_\varepsilon x - Tx\| < \varepsilon$ for all $x \in C$ and $\dim(Im(T_\varepsilon)) < \infty$.

Let us see some examples that illustrate the use of Schauder's theorem.

Example 4.2. Consider the initial value problem

$$\begin{cases} x'(t) = f(t, x(t)), \\ x(t_0) = x_0, \end{cases}$$

with $f : \Omega \subset \mathbb{R} \times \mathbb{R}^n \to \mathbb{R}^n$ continuous. As seen in Exercise 2.5, it is still possible to prove the existence of a (not necessarily unique) solution in a neighborhood of t_0 without the Lipschitz assumption. But now we may give a very direct proof using Schauder's theorem.

For fixed $\delta, R > 0$, let $E = C([t_0 - \delta, t_0 + \delta], \mathbb{R}^n)$, and consider the convex, bounded, and closed $C \subset E$ given by

$$C = \{x \in E : \|x - x_0\|_\infty \leq R\}.$$

We claim that if δ is small enough, then the fixed point operator

$$Tx(t) = x_0 + \int_{t_0}^{t} f(s, x(s))\, ds$$

is well defined, and $T(C) \subset C$.

Indeed, fix $\hat{\delta}, r > 0$ such that $K := [t_0 - \hat{\delta}, t_0 + \hat{\delta}] \times \overline{B_R(x_0)} \subset \Omega$; then there exists M such that $|f(t, x)| \le M$ for $(t, x) \in K$. Take $\delta \le \hat{\delta}$; then for $x \in C$ it follows that $(t, x(t)) \in K$ and

$$|Tx(t) - x_0| = \left| \int_{t_0}^{t} f(s, x(s))\, ds \right| \le \delta M.$$

Thus, it suffices to choose $\delta \le \min\{\hat{\delta}, \frac{R}{M}\}$. It remains to prove that T is continuous and $\overline{T(C)}$ is compact.

Continuity is clear since T is an integral operator. For example, we may take $x_n \to x$ in C; then $f(\cdot, x_n) \to f(\cdot, x)$ uniformly and

$$|Tx_n(t) - Tx(t)| = \left| \int_{t_0}^{t} f(s, x_n) - f(s, x)\, ds \right| \le \delta \|f(\cdot, x_n) - f(\cdot, x)\|_\infty \to 0$$

as $n \to \infty$. Since the right-hand side does not depend on t, the result is proven.

Finally, compactness is an immediate consequence of the Arzelá–Ascoli theorem and the fact that

$$|Tx(t + h) - Tx(t)| = \left| \int_{t}^{t+h} f(s, x)\, ds \right| \le |h| M.$$

Example 4.3. (Abstract nonresonant problem) Let E and F be Banach spaces, and assume that the linear operator $L : D \subset E \to F$ has an inverse $L^{-1} : F \to E$. If $N : E \to F$ is a continuous operator, then the problem

$$Lu = Nu$$

may be written as the fixed point problem $u = L^{-1}Nu$. As mentioned, in many applications L^{-1} is compact, so the operator $T := L^{-1}N$ is also compact if we assume that N maps bounded sets into bounded sets. Indeed, T is continuous because it is the composition of two continuous operators; moreover, if $C \subset E$ is bounded, then $N(C)$ is bounded, and hence $\overline{L^{-1}(N(C))}$ is compact.

Remark 4.2. Observe that if $dim(E) = \infty$, then L cannot be continuous since it has a compact inverse.

In particular, the previous assumptions are satisfied in the case of the problem

$$u''(t) = f(t, u(t)), \qquad u(0) = u(1) = 0, \tag{4.1}$$

with

$$Lu := u'', \qquad Nu := f(\cdot, u),$$
$$E = F = C([0,1]), \qquad D = \{u \in C^2([0,1]) : u(0) = u(1) = 0\}.$$

The fact that L has an algebraic inverse was mentioned in previous chapters; furthermore, we have proven that if we consider D equipped with the C^2 norm, then $L^{-1} : E \to D$ is continuous. We claim that L^{-1}, now considered as an endomorphism of E, is compact. This is simply a consequence of the compactness of the embedding $D \hookrightarrow E$ and can also be directly seen as follows. Let $\varphi \in C([0,1])$, and let $u = L^{-1}\varphi$; then $\|u\|_\infty \leq \frac{1}{2}\|u'\|_\infty \leq \frac{1}{2}\|u''\|_\infty = \frac{1}{2}\|\varphi\|_\infty$. This proves that L^{-1} is continuous; furthermore, if $\|\varphi\|_\infty \leq R$, then, for arbitrary $s,t \in [0,1]$,

$$|u(t) - u(s)| = |u'(\theta)(t-s)| \leq R|t-s|.$$

This shows that $L^{-1}(\overline{B_R(0)})$ is bounded and equicontinuous, so by the Arzelá–Ascoli theorem we conclude that its closure is compact. Hence, L^{-1} is a compact operator. On the other hand, the continuity of N is easily verified, and if $\|\varphi\|_\infty \leq R$, then

$$\|N\varphi\|_\infty \leq \max_{(t,u)\in[0,1]\times[-R,R]} |f(t,u)|.$$

Thus, N maps bounded sets into bounded sets.

An immediate existence result for the abstract case is obtained when the range of N is a bounded set or, more generally, when N has linear growth, i.e.,

$$\|Nu\| \leq A\|u\| + B$$

for some constants A, B. Indeed, because L^{-1} is continuous, there exists a constant c such that $\|L^{-1}\varphi\| \leq c\|\varphi\|$ for all φ; hence

$$\|L^{-1}Nu\| \leq c\|Nu\| \leq c(A\|u\| + B).$$

Assume that $cA < 1$, and let $R := \frac{cB}{1-cA}$. If $\|u\| \leq R$, then

$$\|Tu\| \leq cAR + cB = R.$$

Thus, $T(\overline{B_R(0)}) \subset \overline{B_R(0)}$, and by Schauder's theorem we deduce that T has a fixed point. This implies that problem (4.1) admits solutions when $|f(t,u)| \leq A|u| + B$ with $A < 4$. The sharper condition $A < \pi^2$ is obtained if, instead, one considers $E = F = L^2(0,1)$. Compare with Example 1.3 and Remark 1.1.

Example 4.4. (Hartman condition) Let $f : [0,1] \times \mathbb{R}^n \to \mathbb{R}^n$ be continuous, and assume there exists $R > 0$ such that

$$f(t,u) \cdot u \geq 0 \qquad (4.2)$$

for all $t \in [0,1]$ and all $u \in \mathbb{R}^n$ such that $|u| = R$. For $n = 2$, it was proven in Exercise 1.7 that if the previous inequality is strict and f is C^1 with respect to u, then the Dirichlet problem

$$u''(t) = f(t, u(t)) \qquad u(0) = u_0, \ u(1) = u_1$$

has a solution for arbitrary boundary data u_0, u_1 such that $|u_0|, |u_1| \le R$. Using an approximation argument it is proven (Exercise 1.9) that the conclusion is still true when f is only continuous and satisfies (4.2); thus, the result is easily verified for arbitrary n by the general version of Brouwer's theorem.

However, Schauder's theorem makes possible a very simple and direct proof for the strict case. Let $\lambda > 0$, and define $P : \mathbb{R}^n \to \mathbb{R}^n$ by

$$P(u) = \begin{cases} u & \text{if } |u| \le R, \\ R\frac{u}{|u|} & \text{if } |u| > R. \end{cases}$$

The equation

$$u''(t) - \lambda u(t) = f(t, P(u(t))) - \lambda P(u(t))$$

has a bounded right-hand side, so by Schauder's theorem it has at least one solution u satisfying the Dirichlet boundary conditions. Next, define $\phi(t) := \frac{|u(t)|^2}{2}$; then

$$\phi'(t) = u(t) \cdot u'(t), \qquad \phi''(t) = u(t) \cdot u''(t) + |u'(t)|^2.$$

If ϕ achieves its absolute maximum at some t_0 with $\phi(t_0) > \frac{R^2}{2}$, then $t_0 \in (0, 1)$ and

$$0 \ge \phi''(t_0) \ge u(t_0) \cdot u''(t_0) = u(t_0) \cdot f\left(t_0, R\frac{u(t_0)}{|u(t_0)|}\right) > 0,$$

a contradiction. Thus $\|u\|_\infty \le R$, so we conclude that u solves the original problem. As shown in Chap. 1, the nonstrict case follows taking $f_k(t, u) := f(t, u) + \frac{u}{k}$ and applying the Arzelá–Ascoli theorem to a sequence $\{u_k\}$ of solutions for f_k such that $\|u_k\|_\infty \le R$. Analogous results are easily obtained for periodic or Neumann boundary conditions.

4.3 Upper and Lower Solutions Strike Again

In Chap. 3, we proved that certain problems admit solutions when an ordered couple $\alpha \le \beta$ of a lower and an upper solution exist. The proof was based on monotone iterations and required, for some appropriate $\lambda > 0$ and all fixed t, the function $f(t, u) - \lambda u$ to be nonincreasing for $u \in [\alpha(t), \beta(t)]$. Now we have a very powerful tool for finding fixed points, so it is worth trying to apply it to this situation. If we consider problem (4.1), then it is an exercise for the reader to verify that the operator given by $Tv := u$, where u is the unique solution of the problem $u''(t) - \lambda u(t) = f(t, v(t)) - \lambda v(t)$, is compact and maps the closed, convex, and bounded set $C := \{u \in C([0,1]) : \alpha \leq u \leq \beta\}$ into itself.

However, this proof is not valid for arbitrary continuous f since such a constant λ may not exist. In this section, we shall demonstrate that Schauder's theorem allows us to extend the result for general f, which, furthermore, may also depend on u'. But this new extension will come later, once we understand how the method works. So let us solve (4.1) first.

To this end, we shall apply, as was previously done, a truncation method. Assume that a lower and an upper solution $\alpha \leq \beta$ exist, and define $P : [0,1] \times \mathbb{R} \to \mathbb{R}$ by

$$P(t,u) = \begin{cases} u & \text{if } \alpha(t) \leq u \leq \beta(t), \\ \alpha(t) & \text{if } \alpha(t) > u, \\ \beta(t) & \text{if } u > \beta(t). \end{cases}$$

Next, fix any constant $\lambda > 0$, and consider the problem

$$u''(t) - \lambda u(t) = f(t, P(t, u(t))) - \lambda P(t, u(t)), \qquad u(0) = u(1) = 0. \qquad (4.3)$$

Since the right-hand side of (4.3) is bounded, we know from the previous section that the problem has at least one solution u. Moreover, if we suppose, for example, $u \not\leq \beta$, then $u - \beta$ achieves an absolute positive maximum at some value $t_0 \in (0,1)$. In particular, $P(t_0, u(t_0)) = \beta(t_0)$, so

$$u''(t_0) - \lambda u(t_0) = f(t_0, \beta(t_0)) - \lambda \beta(t_0) \geq \beta''(t_0) - \lambda \beta(t_0),$$

and hence $(u - \beta)''(t_0) \geq \lambda u(t_0) - \beta(t_0) > 0$, a contradiction. In the same way, it is seen that $u \geq \alpha$, and hence u is a solution of the original problem.

4.3.1 Bounds for the Derivative: Nagumo Condition

In view of the conclusions of the last section, we might attempt to apply the method of upper and lower solutions to a more general problem:

$$\begin{cases} u''(t) = f(t, u(t), u'(t)), \\ u(0) = u(1) = 0, \end{cases} \qquad (4.4)$$

with $f : [0,1] \times \mathbb{R}^2 \to \mathbb{R}$ continuous. As before, we shall assume that $\alpha \leq \beta$ are C^2 functions satisfying

$$\alpha''(t) \geq f(t, \alpha(t), \alpha'(t)), \qquad \beta''(t) \leq f(t, \beta(t), \beta'(t)),$$

$$\alpha(0) \leq u_0 \leq \beta(0), \qquad \alpha(1) \leq u_1 \leq \beta(1).$$

Very soon it becomes clear that if we want to define a fixed point operator as before, we need to work in a space that admits first-order derivatives, e.g., $C^1([0,1])$. But then we need a closed, bounded, and convex invariant region in this new space, and this requires obtaining estimates on the derivatives.

Before clarifying this point, let us give an example showing that the presence of an ordered couple $\alpha \le \beta$ is not enough to guarantee the existence of solutions. This fact had already been observed in [90] and made more general by Habets and Pouso in [43].

Consider the problem

$$\left(\frac{u'}{\sqrt{1+u'^2}}\right)'(t) = u(t) + 2$$

with the Dirichlet condition

$$u(0) = u(T) = 0.$$

It is readily seen that $\alpha \equiv -3$ and $\beta \equiv 3$ are respectively a lower and an upper solution. However, the problem has no solutions for $T \ge \sqrt{2}$.

Indeed, multiply both sides of the equation by $u'(t)$ and integrate to verify that any solution of the problem must have constant energy

$$E \equiv \frac{1}{\sqrt{1+u'(t)^2}} + \frac{(u(t)+2)^2}{2}.$$

From the Dirichlet condition we deduce that $E > 2$; in particular, this implies that u cannot take the value -2, and hence $u(t) > -2$ for all t. Next, fix t_0 such that $u(t_0) = \min_{0 \le t \le T} u(t)$; then, for $t \ge t_0$ we may set $s := u(t)$ and use the previous equality to obtain

$$t - t_0 = \int_{t_0}^{t} dt = \int_{u(t_0)}^{u(t)} \frac{E - \frac{(s+2)^2}{2}}{\sqrt{1 - \left(E - \frac{(s+2)^2}{2}\right)^2}} ds.$$

Now take $w = w(s) = E - \frac{(s+2)^2}{2}$; then $0 < w < 1$ and $(s+2)^2 = 2(E-w) > 2$, and, thus,

$$t - t_0 = \int_{w(u(t_0))}^{w(u(t))} \frac{w}{\sqrt{2(E-w)}\sqrt{1-w^2}} dw < \frac{1}{\sqrt{2}} \int_{0}^{1} \frac{w}{\sqrt{1-w^2}} dw = \frac{1}{\sqrt{2}}.$$

In the same way, for $t \le t_0$ it is seen that $t_0 - t \le \frac{1}{\sqrt{2}}$, and hence $T < \frac{2}{\sqrt{2}} = \sqrt{2}$.

It is worth mentioning that the operator $\left(\frac{u'}{\sqrt{1+u'^2}}\right)'$ is well known in the literature as the *mean curvature* operator, which corresponds, in its general version, to the equation for a hypersurface defined by the graph of a function $u : \Omega \subset \mathbb{R}^n \to \mathbb{R}$ with mean curvature H:

$$\text{div}\left(\frac{\nabla u}{\sqrt{1+\nabla u^2}}\right) = nH.$$

At first glance, it is not obvious why solutions of the previous problem do not exist for large values of T, although we may observe, in the first place, that the equation is equivalent to

$$u''(t) = (u(t) + 2)\left(1 + u'(t)^2\right)^{3/2}.$$

In other words, the nonlinearity has *cubic growth* with respect to u'.

In contrast with this case we observe that if f has, for example, *sublinear* growth with respect to u', then the existence of solutions follows almost exactly as in the previous cases: for fixed $v \in C^1([0,1])$, let $u = Tv$ be the unique solution of the linear problem

$$u''(t) - \lambda u(t) = f(t, P(t, v(t)), v'(t)) - \lambda P(t, v(t)), \qquad u(0) = u(1) = 0.$$

It is easy to check that T is compact and, due to sublinearity,

$$\|u\|_{C^1} \le c\|u''(t) - \lambda u(t)\|_\infty \le A\|v'\|_\infty + B \le A\|v\|_{C^1} + B$$

for some constant $A < 1$. Taking $R = \frac{B}{1-A}$, it follows that $T(\overline{B_R(0)}) \subset \overline{B_R(0)}$, and the result follows by Schauder's theorem.

We are still far from solving the most general situation; nonetheless, for the moment we have enough evidence to conclude that the growth of the nonlinearity plays a crucial role: if f grows slowly with respect to u', then lower and upper solutions ensure the existence of solutions; if it grows too fast, then solutions may not exist. The sufficient condition that shall be introduced in what follows permits f to grow quadratically with respect to u'.

Inspired by the case $f = f(t, u)$, we may first observe that, in general, when no restrictions are assumed, it is not possible to obtain bounds for u'. Thus, it shall be convenient to use a truncation function for the u' coordinate too. Let $R > 0$ to be established, and define

$$Q(v) = \begin{cases} v & \text{if } |v| \le R, \\ R & \text{if } v > R, \\ -R & \text{if } v < -R. \end{cases}$$

Next, we may set $\lambda > 0$ and consider the problem

$$u''(t) - \lambda u(t) = f(t, P(t, u(t)), Q(u'(t))) - \lambda P(t, u(t)), \qquad u(0) = u(1) = 0.$$

Again, because the right-hand side is bounded, applying Schauder's theorem now in $C^1([0,1])$ we deduce the existence of a solution u. We shall impose additional conditions in order to guarantee that $\alpha \le u \le \beta$ and $\|u'\|_\infty \le R$, so u is a solution of the original problem.

Let us begin exactly as we did in the previous case: if, for example, an absolute positive maximum of the function $u - \beta$ is achieved at some $t_0 \in (0,1)$, then $P(t_0, u(t_0)) = \beta(t_0)$. However, we cannot yet use the fact that β is an upper solution unless we are able to verify that $Q((u'(t_0)) = \beta'(t_0)$. Since t_0 is a critical point of $u - \beta$, we know that $u'(t_0) = \beta'(t_0)$, so all our difficulties are overcome if we choose R satisfying $R \geq \|\beta'\|_\infty$: in this case, $Q((u'(t_0)) = Q(\beta'(t_0)) = \beta'(t_0)$. Analogous considerations for the remaining case also impose the condition $R \geq \|\alpha'\|_\infty$; summarizing, if we take $R \geq \max\{\|\alpha'\|_\infty, \|\beta'\|_\infty\}$, then $\alpha \leq u \leq \beta$.

As a consequence, we deduce that $u''(t) = f(t, u(t), Q(u'(t)))$. Now we want to find a condition that will allow us to prove that $\|u'\|_\infty \leq R$. Although there exist several variants, the most famous is the one known in the literature as the *Nagumo condition* [89]. Let

$$E = \{(t, u, v) : t \in [0, 1], \alpha(t) \leq u \leq \beta(t), |v| \leq R\}, \tag{4.5}$$

and assume, for all $(t, u, v) \in E$, that

$$|f(t, u, v)| \leq \psi(|v|), \tag{4.6}$$

where $\psi : [0, +\infty) \to (0, \infty)$ satisfies

$$\int_0^{+\infty} \frac{s}{\psi(s)} \, ds = \infty. \tag{4.7}$$

For example, taking $\psi(s) = As^2 + B$ we observe that quadratic growth is admitted.

We claim that if u is a solution of the previous problem, then $\|u'\|_\infty < R$. Indeed, by Rolle's theorem we may fix $t_0 \in (0, 1)$ such that $u'(t_0) = 0$. If, for example, $u'(t) \geq R$ for some t, then there exists t_1 such that $u(t_1) = R$ and $0 < u'(t) < R$ for t between t_0 and t_1. Using the substitution $s = u'(t)$ we obtain

$$\int_0^R \frac{s}{\psi(s)} \, ds = \int_{t_0}^{t_1} \frac{u'(t)u''(t)}{\psi(u'(t))} \, dt = \int_{t_0}^{t_1} \frac{u'(t)f(t, u(t), Q(u'(t)))}{\psi(u'(t))} \, dt.$$

Since $(t, u(t), Q(u'(t))) = (t, u(t), u'(t)) \in E$ and $u'(t) > 0$, we conclude from (4.6) that

$$\int_0^R \frac{s}{\psi(s)} \, ds \leq \left| \int_{t_0}^{t_1} u'(t) \, dt \right| = |u(t_1) - u(t_0)| \leq 2 \max\{\|\alpha\|_\infty, \|\beta\|_\infty\}.$$

Taking into account (4.7), this implies that R cannot be too large. Thus, we have proven the following propsition.

Proposition 4.1. *Let $\alpha \leq \beta$ be respectively a lower and an upper solution of (4.4), let $R > \max\{\|\alpha'\|_\infty, \|\beta'\|_\infty\}$, and let E be as before. Suppose that f satisfies (4.6) over E for some ψ such that*

$$\int_0^R \frac{s}{\psi(s)} \, ds > 2 \max\{\|\alpha\|_\infty, \|\beta\|_\infty\}. \tag{4.8}$$

Then every solution of the problem

$$u''(t) = f(t, u(t), Q(u'(t))), \qquad u(0) = u(1) = 0,$$

with $\alpha \leq u \leq \beta$, verifies $\|u'\|_\infty < R$.

Remark 4.3. The same conclusions are obtained for other boundary conditions such as periodic or homogeneous Neumann conditions. For the nonhomogeneous Dirichlet condition $u(0) = u_0$, $u(1) = u_1$, the result is analogous, but taking the integral in (4.8) $[r,R]$ instead of $[0,R]$, where $r := |u_1 - u_0| < R$. This is due to the fact that Rolle's theorem does not apply and one must employ, instead, Lagrange's theorem. The same condition serves for the nonhomogeneous Neumann condition $u'(0) = v_0$, $u'(1) = v_1$, with $r = \max\{|v_0|, |v_1|\}$.

More generally, if u is a solution of the equation $u''(t) = f(t, u(t), Q(u'(t)))$ such that $\alpha \le u \le \beta$ and no boundary conditions are assumed, then the previous Nagumo condition with $R > r := \max\{\beta(1) - \alpha(0), \alpha(1) - \beta(0)\}$ implies that $\|u'\|_\infty < R$. An alternative Nagumo condition is given in Exercise 4.11. A direct proof of the result when f grows quadratically is proposed in Exercise 4.12.

Corollary 4.1. *Let* $\alpha \le \beta$, R, *and* f *be as before. Then* (4.4) *has at least one solution* u *such that* $\alpha \le u \le \beta$ *and* $\|u'\|_\infty < R$.

4.3.2 Periodic Solutions of Pendulum Equation

In this section, we give an application of the method of upper and lower solutions to the problem of finding periodic solutions of the forced pendulum equation with friction, namely,

$$u''(t) + au'(t) + \sin u(t) = p(t),$$

$$u(0) = u(1), \quad u'(0) = u'(1).$$

For simplicity, we shall assume that p is a continuous function. As already mentioned, solutions may not exist if the average of p is large since integrating at both sides we obtain

$$\int_0^1 \sin u(t)\, dt = \int_0^T p(t)\, dt = \bar{p}.$$

Thus, a necessary condition for the existence of solutions is that $-1 \le \bar{p} \le 1$. On the other hand, a well-known result establishes (by variational methods) that if $a = 0$, then the problem has a solution when $\bar{p} = 0$.

Remark 4.4. Note that if u is a solution, then $u + 2k\pi$ is also a solution for arbitrary $k \in \mathbb{Z}$. Thus *a solution* means, in fact, infinitely many solutions. However, solutions obtained from one another by adding a multiple of 2π are essentially the same, so it is convenient to distinguish as *geometrically different* those solutions that do not differ by a multiple of 2π.

The previous comments motivate us to write the forcing term p as $p_0 + c$, where $p_0 = p - \bar{p}$ has zero average and $c = \bar{p}$ is a constant. Henceforth, we consider the equivalent problem

$$u''(t) + au'(t) + \sin u(t) = p_0(t) + c, \quad u(0) = u(1), \quad u'(0) = u'(1). \quad (4.9)$$

We shall prove the following theorem.

Theorem 4.7. *Let $p_0 \in C([0,1])$ be such that $\overline{p}_0 = 0$. Then there exist numbers $d(p_0)$ and $D(p_0)$, with $-1 \leq d(p_0) \leq D(p_0) \leq 1$, such that (4.9) has solutions if and only if $c \in [d(p_0), D(p_0)]$.*

A slightly more general version of this result for $a = 0$ was proven in [25] by variational methods. The general case was solved in [36], where it is also shown that, if $c \in (d(p_0), D(p_0))$, then there exist at least two geometrically different solutions. As mentioned, for $a = 0$ it is additionally known that $d(p_0) \leq 0 \leq D(p_0)$; a family of functions given in [97] shows that the latter property is not always true for $a \neq 0$. But even if it cannot be asserted that the problem always has solutions when $c = 0$, Theorem 4.7 guarantees, first of all, that the set of all possible values of c such that the problem has solutions is nonempty. When $\|p\|_\infty \leq 1$, this is easily seen by the method of upper and lower solutions (Exercise 3.2); for arbitrary p, it can be deduced in the following way. First, observe as before that if u is a solution, then c must be equal to the average of the function $\sin u(t)$; thus, we may consider, instead, the integrodifferential problem

$$u''(t) + au'(t) + \sin u(t) = p_0(t) + c(u), \qquad u(0) = u(1), \qquad (4.10)$$

where $c(u) := \int_0^1 \sin u(t)\, dt$.

As the reader may have noticed, the advantage of this new setting consists in that only the first periodicity condition is needed since the other one is automatically satisfied. Indeed, if u is a solution of (4.10), then taking the average at both sides yields

$$\int_0^1 u''(t)\, dt + c(u) = c(u),$$

and we conclude from Barrow's rule that $u'(0) = u'(1)$. Summarizing, if $I(p_0)$ denotes the set of all $c \in \mathbb{R}$ such that (4.9) is solvable, then

$$I(p_0) := \{c(u) : u \text{ is a solution of (4.10)}\}.$$

Thus, the existence of solutions of (4.9) is guaranteed by the following lemma.

Lemma 4.1. *For each $r \in \mathbb{R}$ there exists at least one solution of (4.10) such that $u(0) = u(1) = r$.*

Proof. Apply Schauder's theorem to the operator $v \mapsto u$, where u is the unique solution of the problem

$$u''(t) + au'(t) + \sin v(t) = p_0(t) + c(v), \qquad u(0) = u(1) = r.$$

$\square$

From the lemma, for each $r \in \mathbb{R}$ there exists at least one solution u_r of (4.10) that takes the value r at $t = 0, 1$. As mentioned, $u_r'(0) = u_r'(1)$ and, in particular, $I(p_0) \neq \emptyset$ since $c(u_r) \in I(p_0)$.

However, it might happen that $c(u_r)$ has the same value for all $r \in \mathbb{R}$ and $I(p_0)$ reduces to just one point. It is not yet known whether a function such that $I(p_0)$ is a singleton exists or not; the problem, if such a situation occurs, is called *degenerate*. It has been proven that the set of functions of zero average for which a problem is nondegenerate is open and dense in $\tilde{C}([0, 1])$, the space of zero-average continuous functions (e.g., [86]), but the general problem is still open. From the previous lemma it is deduced that, if $I(p_0) = \{c\}$, then the periodic solutions of (4.9) form a *continuum* since, in that case, $c(u_r) = c$ for every solution u_r of (4.10). More generally, it was proven in [98] that degeneracy is equivalent to any of the following statements:

1. For each $r \in \mathbb{R}$ there exists a unique solution u_r such that $u_r(0) = r$.
2. There exists a continuous path $r \mapsto u_r$ such that

$$\lim_{r \to \pm\infty} u_r(t) = \pm\infty$$

uniformly in t.

In some cases, it is easy to show that $I(p_0)$ has a nonempty interior: for example, if $\|p_0\|_\infty < 1$, then there is a neighborhood I of 0 such that $\|p_0 + c\|_\infty \le 1$ for $c \in I$, and hence, using the upper and lower solutions method, it follows that $I \subset I(p_0)$. Another example shall be given in Exercise 4.10. On the other hand, cases of degenerate equations apparently (but only apparently!) similar to the pendulum equation are known: in [13] it was proven that the problem

$$u''(t) + \sin(t + u(t)) = c, \qquad u(0) = u(2\pi), \ u'(0) = u'(2\pi)$$

has no solutions for $c \ne 0$ and a continuum of solutions when $c = 0$.

Now let us return to Theorem 4.7. From the previous considerations, we know that $I(p_0)$ is nonempty and bounded, so it remains to see that $I(p_0)$ is connected and closed.

Assume that $c_1, c_2 \in I(p_0)$ are such that $c_1 < c_2$, and let u_1 and u_2 be solutions for c_1 and c_2, respectively. Then for any $c \in (c_1, c_2)$ it is seen that

$$u_1''(t) + au_1'(t) + \sin u_1(t) < p_0(t) + c < u_2''(t) + au_2'(t) + \sin u_2(t),$$

and thus u_1 and u_2 are respectively an upper and a lower solution of the problem. The difficulty is that we do not know whether or not they are well ordered, so the method cannot be applied yet. But this is our lucky day: by periodicity, it is clear that u_1 may be replaced by $u_1 + 2k\pi$, where the integer k is chosen in such a way that $u_2 \le u_1 + 2k\pi$. Hence, problem (4.9) has a solution, and we conclude that $c \in I(p_0)$.

Remark 4.5. As shown in [36], using degree theory (Chap. 5) it is possible to prove, for such c, the existence of a second solution, that is, one that is geometrically different from the previous one.

Finally, suppose that a sequence $\{c_n\} \subset I(p_0)$ converges to some c, and take a solution u_n of (4.10) such that $u_n(0) = u_n(1) = r_n$ and $c(u_n) = c_n$. By periodicity, we may assume that $r_n \in [0, 2\pi]$.

As we proved in similar problems, there exists a constant C such that

$$\|u_n - r_n\|_\infty \le \frac{1}{2}\|u_n'\|_\infty \le C\|u_n'' + au_n'\|_\infty \le C(\|p_0\|_\infty + 2).$$

In particular, $\|u_n'\|_\infty$ is bounded, and hence $\|u_n''\|_\infty$ is also bounded. By the Arzelá–Ascoli theorem there exists a subsequence (still denoted $\{u_n\}$) that converges to some u for the C^1 norm. It is clear that $c(u_n) \to c(u)$, and from the equation

$$u_n'' + au_n' + \sin u_n = p_0 + c(u_n)$$

it is seen that u_n'' converges uniformly to some continuous function w. Writing $u_n'(t) = u_n'(0) + \int_0^t u_n''(s)\,ds$, we conclude that $u'' = w$ and that u solves (4.10) with $c(u) = c$. This shows that $I(p_0)$ is closed and so completes the proof.

To end this section, it is worth mentioning that the quantities $d(p_0)$ and $D(p_0)$ in Theorem 4.7, regarded as functions from $\tilde{C}([0,1])$ to $\mathbb{R}$, are continuous. We sketch a simple proof of this fact; details are left to the reader.

Let p_0^n be continuous with zero average, and assume that $p_0^n \to p_0$ uniformly. If $c_n \in I(p_0^n) \subset [-1,1]$, then take u_n such that

$$u_n''(t) + au_n'(t) + \sin u_n(t) = p_0^n(t) + c_n, \qquad u_n(0) = u_n(1), \quad u_n'(0) = u_n'(1)$$

and $u_n(0) \in [0, 2\pi]$. Thus $\{u_n\}$ is bounded for the C^2 norm, so taking a subsequence we may assume that c_n converges to some $c \in [-1, 1]$ and u_n converges for the C^1 norm to some $u \in C^1([0,1])$. Clearly, u satisfies the periodic conditions, and it is readily seen that

$$u''(t) + au'(t) + \sin u(t) = p_0(t) + c,$$

so $c \in I(p_0)$. This proves that $D(p_0) \ge \limsup_{n \to \infty} D(p_0^n)$.

We claim that, conversely, $D(p_0) \le \liminf_{n \to \infty} D(p_0^n)$. Indeed, otherwise there exists a subsequence (still denoted p_0^n) such that $c_n := D(p_0^n) \to c_* < D(p_0)$. Set u_n as before and take a solution u of (4.9) for $c = D(p_0)$. Fix $\delta \in (0, D(p_0) - c_*)$; then for large enough n we have

$$u''(t) + au'(t) + \sin u(t) > p_0^n(t) + c_* + \delta.$$

On the other hand, when n is large we also have

$$u_n''(t) + au_n'(t) + \sin u_n(t) = p_0^n(t) + c_n < p_0^n(t) + c_* + \delta.$$

Then u is a lower solution and u_n is an upper solution of the problem for $p_0^n + c_* + \delta$, so replacing u_n by $u_n + 2k\pi$, if necessary, we deduce the existence of a solution between u and u_n. Hence $c_* + \delta \in I(p_0^n)$ and, in particular, $D(p_0^n) \ge c_* + \delta$, a contradiction when n is large. We conclude that $D(p_0^n) \to D(p_0)$ and, analogously, that $d(p_0^n) \to d(p_0)$.

4.4 A First Glimpse at the Leray–Schauder Continuation Technique

As has been shown since the outset of the book, when fixed point methods are applied to different boundary value problems, it proves to be very helpful to have a *priori bounds* of the solutions. In some sense, getting a fixed point of a certain problem is like looking for a needle in a haystack; a priori bounds allow us to make the haystack smaller. There are situations in which bounds can be obtained uniformly with respect to a certain parameter: for example, the equation

$$u''(t) = \lambda f(t, u(t))$$

under homogeneous Dirichlet conditions and $\lambda \in [0,1]$ admits, when f is nondecreasing in u, a priori bounds independent of λ. In Chap. 2, this fact was used together with the implicit function theorem to prove that solutions can be "continued" up to the value $\lambda = 1$, provided that f is smooth. The next result shows that the existence of uniform bounds is enough by itself to ensure that solutions exist, regardless of the smoothness assumption.

Theorem 4.8. *(Schaefer) Let E be a Banach space and $T : E \to E$ a compact operator. Assume there exists $R > 0$ such that $\|x\| < R$ for any $x \in E$ such that $x = \lambda T x$ for some $\lambda \in (0,1]$. Then T has at least one fixed point.*

Proof. Define the operator $T^* : E \to E$ given by

$$T^* x = \begin{cases} Tx & \text{if } \|Tx\| \leq R, \\ \frac{R}{\|Tx\|} Tx & \text{if } \|Tx\| > R. \end{cases}$$

It is readily seen that T^* is compact; furthermore, its range is contained in $\overline{B_R(0)}$, so, in particular, it has a fixed point $x \in \overline{B_R(0)}$. If $\|Tx\| > R$, then

$$x = T^* x = \frac{R}{\|Tx\|} Tx,$$

so $\|x\| = R$ and $x = \lambda T x$ with $\lambda = \frac{R}{\|Tx\|} < 1$, a contradiction. Thus, $\|Tx\| \leq R$, and hence x is a fixed point of T. $\square$

Example 4.5. Consider the nonhomogeneous Liénard equation

$$u''(t) + f(u(t))u'(t) + u(t) = p(t),$$

with $f : \mathbb{R} \to \mathbb{R}$ and $p : [0,T] \to \mathbb{R}$ continuous, under T-periodic conditions

$$u(0) - u(T) = u'(0) - u'(T) = 0.$$

We shall prove the existence of solutions for $T < 2\pi$.

To this end, let $X = \{u \in C^1([0,T]) : u(0) = u(T), u'(0) = u'(T)\}$, and observe that the operator $L : X \cap C^2([0,T]) \to C([0,T])$ given by $Lu := u'' + u$ is invertible. Hence, it is verified that the mapping $T : X \to X$ given by $Tv = u$, where $u \in X$ is the unique solution of

$$u''(t) + u(t) = p(t) - f(v(t))v'(t), \qquad u(0) = u(T), \quad u'(0) = u'(T),$$

is well defined. By Schaefer's theorem, it suffices to prove the existence of a constant R such that $\|u\|_{C^1} \leq R$ for all u satisfying $u = \lambda Tu$ with $\lambda \in (0,1)$. The latter equality is equivalent to

$$u''(t) + u(t) = \lambda(p(t) - f(u(t))u'(t))$$

and, due to the periodic conditions, taking the average at both sides yields

$$\bar{u} = \lambda\bar{p} - \frac{\lambda}{T}\int_0^T f(u(t))u'(t)\,dt.$$

Let $F(u) := \int_0^u f(s)\,ds$; then $\int_0^T f(u(t))u'(t)\,dt = F \circ u\Big|_0^T = 0$, and hence $\bar{u} = \lambda\bar{p}$. Furthermore, multiplying the equation by $u - \bar{u}$ we obtain

$$\int_0^T (u''(t) + u(t))(u(t) - \bar{u})\,dt = \int_0^T \lambda[p(t) - f(u(t))u'(t)](u(t) - \bar{u})\,dt.$$

Now observe that

$$\int_0^T u(t)(u(t) - \bar{u})\,dt = \int_0^T (u(t) - \bar{u})^2\,dt$$

and

$$\int_0^T f(u(t))u'(t)(u(t) - \bar{u})\,dt = (u - \bar{u})F \circ u\Big|_0^T - \int_0^T F(u(t))u'(t)\,dt = 0,$$

so

$$\int_0^T u'(t)^2\,dt - \int_0^T (u(t) - \bar{u})^2\,dt = -\lambda\int_0^T p(t)(u(t) - \bar{u})\,dt \leq \|p\|_{L^2}\|u - \bar{u}\|_{L^2}.$$

Now, using the Wirtinger inequality $\|u - \bar{u}\|_{L^2} \leq \frac{T}{2\pi}\|u'\|_{L^2}$ (appendix, Sect. B.2.2), we conclude that

$$\|u'\|_{L^2} \leq \frac{2\pi}{2\pi - T}\|p\|_{L^2}.$$

Together with the inequality $|\bar{u}| \leq |\bar{p}|$, this implies that $\|u\|_\infty \leq M$ for some constant M, and from the equation it follows that $\|u''\|_{L^2} \leq N$ for some constant N. Moreover, we know that $\|u'\|_\infty \leq \frac{\sqrt{T}}{2}\|u''\|_{L^2}$, so there exists R such that $\|u\|_{C^1} \leq R$, and this completes the proof.

The same result is obtained for a system

$$u''(t) + Hg(u(t))u'(t) + u(t) = p(t),$$

where Hg denotes the Hessian matrix of a C^2 function $g : \mathbb{R}^n \to \mathbb{R}$.

It is worth mentioning that the result does not hold for the *resonant* case $T = 2\pi$. More shall be said about this problem later on (Exercise 6.6).

Example 4.6. Let us consider the problem of finding T-periodic positive solutions of the delay differential equation

$$x'(t) = -a(t)x(t) + sb(t)g(x(t - \tau(t))), \tag{4.11}$$

where $a, b, \tau : \mathbb{R} \to \mathbb{R}$ are continuous T-periodic positive functions, s is a positive parameter, and $g : [0, +\infty) \to (0, +\infty)$ is continuous, nondecreasing, and superlinear, that is, $\frac{g(x)}{x} \to +\infty$ as $x \to +\infty$.

In the first place, observe that the problem is equivalent to that of finding T-periodic solutions of

$$x'(t) = -a(t)x(t) + sb(t)g(|x(t - \tau(t))|). \tag{4.12}$$

Indeed, it suffices to prove that if x is a T-periodic solution of (4.12), then $x(t) > 0$ for all t. Let $A(t) := \int_0^t a(s)\,ds$, and multiply the equation by $e^{A(t)}$ to obtain

$$\left(e^A x\right)'(t) = se^{A(t)} b(t) g(|x(t - \tau(t))|) > 0.$$

Then, for all t,

$$e^{A(t+T)} x(t + T) > e^{A(t)} x(t),$$

and by periodicity we deduce

$$\left(e^{A(t+T)} - e^{A(t)}\right) x(t) > 0,$$

that is,

$$e^{A(t)} \left(e^{\int_t^{t+T} a(s)\,ds} - 1\right) x(t) > 0.$$

From the positivity of a, we conclude that $x(t) > 0$.

Let $E := \{x \in C(\mathbb{R}, \mathbb{R}) : x(t + T) = x(t) \text{ for all } t\}$; then T-periodic solutions of (4.12) can be obtained as fixed points of the compact operator $\mathscr{T} : E \to E$ given by $\mathscr{T}y = x$, where $x \in E$ is the unique T-periodic solution of the linear equation

$$x'(t) = -a(t)x(t) + sb(t)g(|y(t - \tau(t))|).$$

We claim that if (4.12) has no T-periodic solutions for some s, then it has at least one T-periodic solution for some $\hat{s} < s$. Indeed, by Schaefer's theorem, if $\mathscr{T}$ has no fixed points, then there exist $\lambda_n \in (0, 1)$ and $x_n \in E$ with $\|x_n\| \to \infty$ such that $\lambda_n \mathscr{T} x_n = x_n$ or, equivalently,

$$x_n'(t) + a(t)x_n(t) = \lambda_n s b(t) g(|x_n(t - \tau(t))|).$$

As before, it is proven that $x_n > 0$; thus, x_n is a solution for $s_n := \lambda_n s < s$.

On the other hand, nonexistence for large values of s is immediate since $\frac{g(u)}{u}$ is bounded from below. To verify this, assume that x is a positive T-periodic solution; then $\frac{x'(t)}{x(t)} > -a(t)$ for all t. This implies that $\ln \frac{x(t_1)}{x(t_0)} > -\int_{t_0}^{t_1} a(t)\,dt$ for all $t_0 < t_1$ and, in particular, using the periodicity and the fact that $a > 0$, that $\frac{x_{min}}{x_{max}} \ge e^{-A(T)}$. Hence

$$(\ln x)'(t) + a(t) = s b(t) \frac{g(x(t - \tau(t)))}{x(t - \tau(t))} \frac{x(t - \tau(t))}{x(t)} \ge s b(t) e^{-A(T)} \min_{x \ge 0} \frac{g(x)}{x},$$

so taking the average at both sides, we deduce that $s \bar{b} \min_{x \ge 0} \frac{g(x)}{x} < \bar{a} e^{A(T)}$. Finally, if $\hat{x}$ is a positive T-periodic solution for some $\hat{s}$ and $s \in (0, \hat{s})$, then it is easy to verify that the operator $\mathcal{T}$ maps the set

$$\{x \in X : 0 \le x(t) \le \hat{x}(t) \text{ for all } t\}$$

into itself, and hence, by Schauder's theorem, it has a fixed point (observe, incidentally, that 0 and $\hat{x}$ are respectively a lower and an upper solution of the problem for s). Thus, we have proven the following proposition.

Proposition 4.2. *There exists a constant $s_* > 0$ such that the problem has at least one positive T-periodic solution for $s \in (0, s_*)$ and no positive T-periodic solutions for $s \in (s_*, +\infty)$.*

This result was first established in [110]; the present shorter proof was given in [10]. It is not difficult to prove that if s is small enough, then the problem has at least two positive T-periodic solutions. We shall give a simple proof of this fact in Chap. 6 using degree theory.

Appendix

Brief historical notes. There is a vast literature concerning Brouwer's theorem. According to Dinca and Mawhin [30], its statement was published in 1912 in an article on continuous mappings between manifolds [19]. This paper was a continuation of an earlier work [20], in which it was first proven that open subsets of euclidean spaces with different dimensions cannot be homeomorphic. The no-retraction theorem was proven by Bohl in [17] and independently by Borsuk in [18]. Theorem 4.4 was established by Poincaré in 1884, but it is usually known as the Miranda theorem, after Miranda proved in 1940 its equivalence to Brouwer's theorem (see e.g. [61]). There exists also an infinite-dimensional version in the space l^2, with the unit ball replaced by the *Hilbert cube* (Exercise 5.12).

The method of upper and lower solutions, in the setting of the present chapter, was introduced by Scorza-Dragoni in [105]. The Nagumo condition, as presented

in this chapter, first appeared in [89], although it can be traced back to the works of
Bernstein. It has been generalized in many different ways, both for scalar equations
and systems. For a complete review of the historical and other aspects of the method
of upper and lower solutions, see [28] or [29].

Problems

4.1. Generalize for $n > 2$ the examples and exercises given in Chap. 1.

4.2. Let $p : [0,1] \to \mathbb{R}^n$ be continuous with zero average, and let $g : \mathbb{R}^n \to \mathbb{R}^n$ be
locally Lipschitz, with $x_j g_j(x) \geq 0$ for $|x_j| \geq R_0$ and $\|g\|_\infty < \gamma$ for some $R_0, \gamma > 0$.
Then there exists a constant M depending only on γ and R_0 such that the problem

$$u''(t) + g(u(t)) = p(t), \qquad u(0) = u(1), \quad u'(0) = u'(1)$$

has at least one solution u such that $\|u\|_\infty \leq M$. Deduce that the result is still true
when g is only continuous (see [82]).

    Hint: following the ideas of Sect. 1.3.2, reduce the problem to an equation in $\mathbb{R}^{2n}$
and apply the Poincaré–Miranda theorem. The general case follows by an approximation argument.

4.3. Prove the following particular case of the *Perron–Frobenius theorem*: let $A \in
\mathbb{R}^{n \times n}$ such that $a_{ij} \geq 0$ for all i, j. Then there exist an eigenvalue $\lambda \geq 0$ and an
associated eigenvector x with nonnegative coordinates.

    Hint: consider the set $K := \{x \in \mathbb{R}^n : x_j \geq 0 \text{ for all } j, \sum_{j=1}^n x_j = 1\}$. If $Ax = 0$ for
some $x \in K$, then the result holds with $\lambda = 0$; otherwise, prove that the mapping
$f(x) := \frac{Ax}{\sum_{j=1}^n (Ax)_j}$ has a fixed point in K.

4.4. Let $f : [0,1] \times \mathbb{R} \to \mathbb{R}$ be continuous and bounded. Using Example 4.1, find
a sequence of piecewise linear functions converging uniformly to a solution of the
Dirichlet problem

$$u''(t) = f(t, u(t)), \qquad u(0) = u(1) = 0.$$

4.5. Consider the function $h : L^2(0,1) \times [0,1] \to L^2(0,1)$ defined by $h(f,0) \equiv 1$ and

$$h(f,\lambda) = h_\lambda(f) := \begin{cases} f\left(\frac{t}{\lambda}\right) & \text{if } t < \lambda, \\ 1 & \text{if } t \geq \lambda, \end{cases}$$

when $\lambda > 0$.

1. Prove that h is continuous and vanishes only when $(f,\lambda) = (0,1)$.
2. If $\lambda > 0$, then $\|h(f,\lambda)\|^2 = \lambda(\|f\|^2 - 1) + 1$. In particular, $h_\lambda(B_1(0)) \subset B_1(0)$
   and $h_\lambda(\partial B_1(0)) \subset \partial B_1(0)$.

3. Let $\lambda(f) := \frac{1+\|f\|}{2}$. Prove that $r : \overline{B_1(0)} \to \partial B_1(0)$ given by $r(f) = \frac{h_{\lambda(f)}(f)}{\|h_{\lambda(f)}(f)\|}$ is a continuous retraction.

4. Find a continuous mapping $T : \overline{B_1(0)} \to \overline{B_1(0)}$ with no fixed points.

4.6. Let H be a separable Hilbert space with basis $\{e_n\}_{n \in \mathbb{Z}}$, and consider the isometry $T : H \to H$ defined by

$$T\left(\sum_{n \in \mathbb{Z}} x_n e_n\right) := \sum_{n \in \mathbb{Z}} x_n e_{n+1}.$$

Prove that the function $f : H \to H$ given by

$$f(x) = \frac{1 - \|x\|}{2} e_0 + Tx$$

is continuous and maps the closed unit ball into itself but has no fixed points.

4.7. Let $f : [0,1] \times \mathbb{R}^n \to \mathbb{R}^n$ be continuous and sublinear in x. Prove that the *antiperiodic* problem

$$x'(t) = f(t, x(t)), \qquad x(0) + x(1) = 0$$

has at least one solution.

4.8. Let $c \in C([0,1])$ be nonnegative, and let $f : [0,1] \times \mathbb{R}^{2n} \to \mathbb{R}^n$ be continuous and sublinear in $(u,v) \in \mathbb{R}^{2n}$, that is,

$$\lim_{|(u,v)| \to \infty} \frac{f(t,u,v)}{|(u,v)|} = 0$$

uniformly in t. Prove that the problem $u''(t) - c(t)u(t) = f(t, u(t), u'(t))$ has at least one solution under

1. Dirichlet conditions;
2. Neumann or periodic conditions, provided that $c \not\equiv 0$.

4.9. 1. Extend the method of upper and lower solutions for the scalar equation $u''(t) = f(t, u(t), u'(t))$ under

    a. Nonhomogeneous Dirichlet conditions;
    b. Neumann conditions;
    c. Periodic conditions, assuming that α and β satisfy

$$\alpha(0) - \alpha(T) = \beta(0) - \beta(T) = 0, \qquad \alpha'(0) - \alpha'(T) \geq 0 \geq \beta'(0) - \beta'(T).$$

2. Generalize 1 for a system of equations.

4.10* (Adapted from [36]). In the situation of Sect. 4.3.2, write $u = v + P_0$, where P_0 is the unique solution of the problem

$$P_0''(t) + aP_0'(t) = p_0(t), \qquad P_0(0) = P_0(1), \qquad \overline{P_0} = 0,$$

and verify that $P_0'(0) = P_0'(1)$.

1. Prove that

$$|c(v + P_0) - c(\overline{v} + P_0)| \leq \|v - \overline{v}\|_{L^2}.$$

2. Using the Wirtinger inequality, prove that

$$\|v'\|_{L^2} \leq \frac{1}{2\pi} \|v'' + av'\|_{L^2},$$

and hence

$$|c(v + P_0) - c(\overline{v} + P_0)| \leq \frac{1}{2\pi^2} := M.$$

3. Let $X := \{u \in C^2([0,1]) : u \text{ is a solution of (4.10) with } u(0) = u(1)\}$; then

$$d(p_0) = \min_{u \in X} c(u) = \min_{v + P_0 \in X} c(v + P_0) \leq \inf_{v + P_0 \in X} c(\overline{v} + P_0) + M,$$

and conclude that

$$d(p_0) \leq \inf_{r \in \mathbb{R}} c(r + P_0) + M = M - M(P_0),$$

where

$$M(P_0) = \left[\left(\int_0^1 \cos P_0(t) \, dt \right)^2 + \left(\int_0^1 \sin P_0(t) \, dt \right)^2 \right]^{1/2}.$$

In the same way, deduce that $D(p_0) \geq M(P_0) - M$.

4. Obtain an explicit sufficient condition that guarantees $d(p_0) < 0 < D(p_0)$.

4.11. Prove that the Nagumo condition (4.6) still works if, instead of (4.8), it is assumed that

$$\int_r^R \frac{1}{\psi(s)} ds > 1,$$

with r as in Remark 4.3 and $R > \max\{\|\alpha'\|_\infty, \|\beta'\|_\infty, r\}$.

4.12. Let $f : [0,1] \times \mathbb{R}^2 \to \mathbb{R}$ be continuous, and assume that

$$|f(t, u, v)| \leq c(u)(1 + v^2)$$

for some continuous $c : \mathbb{R} \to \mathbb{R}^+$. Prove that for each $k > 0$ there exists $R = R(k)$ such that if $u''(t) = f(t, u(t), u'(t))$ and $\|u\|_\infty \leq k$, then $\|u'\|_\infty \leq R$.

Hint: if $u' \neq 0$ on (t_0, t_1) and, for example, $u'(t_0) = 0$, then write t as a function of u and define $u_0 = u(t_0)$, $p(u) = u'(t)$ and $q = p^2$. Then

$$\frac{dq}{du} \leq 2c(u)(1 + q(u)), \qquad q(u_0) = 0,$$

and a bound for q is obtained. If u' does not vanish in $[0,1]$, then use Lagrange's theorem.

4.13. Prove that the problem

$$u''(t) + u'(t)^2 = -1, \qquad u(0) = u(T) = 0$$

has no solutions when $T \geq \pi$. Does this contradict the previous exercise?

4.14. Let $f : [0,1] \times \mathbb{R} \to \mathbb{R}$ be continuous and nondecreasing in u.

1. Prove that the semilinear operator $Su := u'' - f(\cdot, u)$ satisfies

$$\|u' - v'\|_{L^2} \leq \frac{1}{\pi} \|Su - Sv\|_{L^2}$$

for all $u, v \in C^2([0,1])$ such that $u(0) = v(0)$ and $u(1) = v(1)$. In particular,

$$\|u\|_\infty \leq \max\{|u(0)|, |u(1)|\} + \frac{1}{2\pi}\left(\|Su\|_\infty + \max_{t \in [0,1]} |f(t, u(0) + t(u(1) - u(0)))|\right).$$

2. Prove that the application $\mathscr{T}_0 : C([0,1]) \to C^2([0,1])$ given by $\mathscr{T}_0 p = u$, where u is the unique solution of the problem $Su = p$ with homogeneous Dirichlet conditions, is well defined and continuous.
3. Let $g : [0,1] \times \mathbb{R} \to \mathbb{R}$ be continuous such that $|g(t,u)| \leq A|u| + B$ with $A < 2\pi$. Prove that the problem

$$u''(t) = f(t, u(t)) + g(t, u(t)), \qquad u(0) = u(1) = 0$$

has at least one solution.
4. Prove that $\mathscr{T} : C([0,1]) \times \mathbb{R}^2 \to C^2([0,1])$ given by $\mathscr{T}(p, u_0, u_T) = u$, where u is the unique solution of the problem $Su = p$ such that $u(0) = u_0, u(1) = u_1$, is well defined and continuous.
5. Prove that, for each $p \in C([0,T])$, the set

$$S^{-1}(p) := \{u \in C^2([0,1]) : u'' + f(t, u(t)) = p(t)\}$$

is homeomorphic to $\mathbb{R}^2$. What can you say about $S^{-1}(p)$ in the linear case $f(t, u) = \varphi(t)u$ with $\varphi \geq 0$?

Chapter 5
Topological Degree: An Introduction

5.1 Brouwer Degree

In this section, we shall introduce the basic aspects of a fundamental topological notion introduced by Brouwer in [19]: the topological degree. Its origins can be traced back to the index of a curve introduced by Cauchy and applied by Poincaré to the study of nonlinear differential equations. Motivated by the study of systems of polynomial equations, Kronecker generalized this idea in 1869 [60]. Although Brouwer's construction used simplicial methods, many other constructions are possible within the framework of algebraic topology or differential geometry. There exist, however, strictly analytical approaches: for example, one was proposed by Nagumo [91, 92] and another one was introduced by Heinz in [49] and inspired by de Rham cohomology. Here, we shall present a simplified version of this latter construction.

But let us start with an intuitive approach. As mentioned in the introduction of the book, the Brouwer degree can be regarded as an "algebraic count" of the zeros of a continuous function $f : \overline{\Omega} \to \mathbb{R}^n$, where Ω is a bounded open subset of $\mathbb{R}^n$. Following the presentation in [83], the idea behind the continuation methods for an equation

$$f(x) = 0$$

consists in embedding it in a family of problems

$$F(x, \lambda) = 0$$

for some continuous $F : \overline{\Omega} \times [0, 1] \to \mathbb{R}^n$ such that $F(x, 1) = f(x)$ for all x and that the equation $F(x, 0) = 0$ has a nonempty set of zeros. If we are lucky enough, at least one of these zeros can be continued up to $\lambda = 1$, thus giving a solution of the original equation.

However, this procedure may fail for several reasons. On the one hand, consider the equation

$$\lambda + (1 - \lambda)x = 0$$

P. Amster, *Topological Methods in the Study of Boundary Value Problems*, Universitext, DOI 10.1007/978-1-4614-8893-4_5, © Springer Science+Business Media New York 2014

over $\Omega = (-1,1)$. When $\lambda = 0$, there exists a unique solution, namely $x = 0$, but as we increase λ, this solution moves continuously to the left until, when $\lambda = \frac{1}{2}$, it "escapes" from the domain through its boundary point $x = -1$. Hence, it becomes clear that we need to prevent solutions from reaching the boundary. On the other hand, the equation

$$x^2 + \lambda - \frac{1}{2} = 0$$

has two zeros in Ω when $\lambda = 0$ that also "disappear" for $\lambda > \frac{1}{2}$. Here, solutions are always away from the boundary: the problem is that the set $\{\pm\frac{\sqrt{2}}{2}\}$ of zeros for $\lambda = 0$ is not *robust*, in a sense that will be specified later.

It is worth noting that the case $n = 2$ is very special since we already have, at least in some situations, an excellent formula for computing zeros. Namely, identify $\mathbb{R}^2$ with $\mathbb{C}$ and assume that $f : \overline{\Omega} \to \mathbb{C}$ is analytic and $\partial\Omega$ can be parameterized by a simple closed curve $\gamma : [a,b] \to \mathbb{C}$ oriented counterclockwise. If f does not vanish over $\partial\Omega$, then the argument principle (or the "zeros and poles theorem") establishes that the number of zeros of f in Ω can be computed by an integral:

$$\#\{z \in \Omega : f(z) = 0\} = \frac{1}{2\pi i}\int_\gamma \frac{f'(z)}{f(z)}\,dz. \tag{5.1}$$

This number is very precise since it provides the *exact* number of zeros. Of course, we count zeros with their multiplicities, so the number might be very large though the function has only one root. However, one thing is clear: if the number is different from zero, then the function vanishes *at least once*. More generally, we might consider an arbitrary point $p \notin f(\partial\Omega)$ and compute the number of solutions of the equation $f(z) = p$. This number will be called the degree of f at p over Ω:

$$deg(f,\Omega,p) := \frac{1}{2\pi i}\int_\gamma \frac{f'(z)}{f(z) - p}\,dz. \tag{5.2}$$

Some evident properties follow from this definition. In the first place, if f is just the identity function, then its degree is 1 or 0 according to whether or not the point belongs to Ω (recall, in this case, that the degree is not defined when $p \in \partial\Omega$):

$$deg(Id,\Omega,p) = \begin{cases} 1 & \text{if } p \in \Omega, \\ 0 & \text{if } p \notin \Omega. \end{cases} \tag{5.3}$$

Moreover, the following translation property is also clear:

$$deg(f,\Omega,p) = deg(f - p,\Omega,0). \tag{5.4}$$

At this stage, it is also trivial that if $deg(f,\Omega,p) \neq 0$, then f takes the value p at least once in Ω. However, after the degree is constructed for arbitrary continuous functions, this will become a fundamental (and not evident) property, called *solution property*. For the moment, observe that if we wish to extend the previous definition for an arbitrary bounded open set Ω, then it is reasonable to assume that the degree is additive over disjoint components. Indeed, this is what happens if we count the number of zeros of a function. Moreover, subsets with no zeros do not contribute to this number; thus, we may condense both ideas in the following property.

If Ω_1 and Ω_2 are disjoint open subsets of Ω and $f \neq p$ over $\overline{\Omega} \setminus (\Omega_1 \cup \Omega_2)$, then

$$deg(f,\Omega,p) = deg(f,\Omega_1,p) + deg(f,\Omega_2,p). \tag{5.5}$$

In particular, (5.5) trivially implies the (also "reasonable") property that the degree with respect to an empty set is zero. The solution property is obtained as well: if f does not vanish in Ω, then take $\Omega_1 = \Omega_2 = \emptyset$ to deduce that $deg(f,\Omega,p) = 0$.

Property (5.5) also allows us to extend the definition given by (5.2) when Ω has a finite number of disjoint "holes" $\overline{\Omega}_1, \ldots, \overline{\Omega}_k$ homeomorphic to discs and, more generally, when Ω is an arbitrary open bounded subset of $\mathbb{C}$. Indeed, as $f \neq p$ on $\partial\Omega$, only a finite number of connected components of Ω will contribute to the degree, and moreover, for each point of the boundary we may remove a small ball without affecting the degree.

Next, we may observe another crucial fact that derives from the preceding definition: two functions that coincide over $\partial\Omega$ necessarily have the same degree. This might seem quite impressive if one merely looks at formula (5.2), which concerns not only f but also f'; however, it becomes completely trivial as soon as one recalls the analytic continuation principle, which in this specific case says that if f and g are analytic and $f = g$ on $\partial\Omega$, then $f \equiv g$. Regardless of this very strong property, it is worth noting that, by definition and the chain rule,

$$\int_{\gamma} \frac{f'(z)}{f(z)-p}\,dz = \int_a^b \frac{f'(\gamma(t))}{f((\gamma(t))-p}\gamma'(t)\,dt = \int_a^b \frac{(f \circ \gamma)'(t)}{f \circ \gamma(t)-p}\,dt$$

$$= \int_{f \circ \gamma} \frac{1}{z-p}\,dz = I(f \circ \gamma, p),$$

where I denotes, as in Chap. 1, the index or winding number of the curve $f \circ \gamma$ with respect to p. On the one hand, this emphasizes the fact that the degree depends only on $f|_{\partial\Omega}$, but on the other hand, it constitutes a valid formula to define the degree when f is just continuous. Indeed, no analyticity is required for defining $I(f \circ \gamma, p)$; also, the curve could be very irregular and the index would still be defined as long as γ is continuous.

Furthermore, the interpretation of the degree as a winding number shows that a more general property holds: the *homotopy invariance*. We shall say that $f, g \rightarrow \mathbb{C}$ continuous such that $p \notin f(\partial\Omega)$ and $p \notin g(\partial\Omega)$ are homotopic (denoted $f \sim g$) if there exists a continuous function $h : \overline{\Omega} \times [0,1] \rightarrow \mathbb{C}$ such that

$$h(z,0) = f(z), \qquad h(z,1) = g(z)$$

for all $z \in \Omega$ and

$$h(z,\lambda) \neq p$$

for all $z \in \partial\Omega$ and $\lambda \in [0,1]$. In this case, it follows trivially that $f \circ \gamma$ and $g \circ \gamma$ are homotopic curves, and hence $I(f \circ \gamma, p) = I(g \circ \gamma, p)$. This yields the *homotopy invariance* property:

$$\text{If} \quad f \sim g, \text{ then} \quad d(f, \Omega, p) = d(g, \Omega, p). \tag{5.6}$$

In particular, (5.6) alone implies (again!) that the degree depends only on the values over the boundary: if $f = g$ on $\partial\Omega$, then $h(z, \lambda) := \lambda g(z) + (1 - \lambda) f(z)$ is a homotopy between f and g.

In what follows, we shall see that the preceding definition may be extended for arbitrary n; more precisely, we shall prove the existence of a mapping (the degree) defined for all open bounded $\Omega \subset \mathbb{R}^n$, all continuous functions $f : \overline{\Omega} \to \mathbb{R}^n$, and all $p \in \mathbb{R}^n$ with $p \notin f(\partial\Omega)$ in such a way that properties (5.3)–(5.6) are satisfied. Furthermore, this can be done in a unique way. The task requires a fair amount of work, so we might start by considering the case $n = 1$ in order to get some inspiration.

Definition of Degree for $n = 1$

From (5.4) we may suppose that $p = 0$; moreover, by additivity we may also assume that $\Omega = (a, b)$. Let $f : [a, b] \to \mathbb{R}$ be continuous, with $f(a), f(b) \neq 0$. We have three cases:

1. $f(a) < 0 < f(b)$. Here, f is homotopic to $g(x) = x - \frac{a+b}{2}$ since g has the same sign of f at a and b, which implies that $\lambda g(x) + (1 - \lambda) f(x)$ does not vanish at those points. Using (5.6), (5.4), and (5.3) we obtain

$$deg(f, \Omega, 0) = deg(g, \Omega, 0) = deg\left(Id, \Omega, \frac{a+b}{2}\right) = 1.$$

2. $f(a)$ and $f(b)$ have the same sign. Here, it suffices to consider g as the segment joining $(a, f(a))$ and $(b, f(b))$. Because g does not vanish in Ω and coincides with f on $\partial\Omega$, we obtain

$$deg(f, \Omega, 0) = deg(g, \Omega, 0) = 0.$$

3. $f(a) > 0 > f(b)$. Take $c > b$ and extend f to a function $\tilde{f} : [a, c] \to \mathbb{R}$ by taking an arbitrary continuous $g : [b, c] \to \mathbb{R}$ such that $g(b) = f(b)$ and $g(c) > 0$. Using (5.5) and the previous cases we obtain

$$0 = deg(\tilde{f}, (a, c), 0) = deg(f, \Omega, 0) + deg(g, (c, d), 0) = deg(f, \Omega, 0) + 1,$$

and thus $deg(f, \Omega, 0) = -1$.

The previous considerations show that the degree can be defined for $n = 1$ in an easy manner, and there is no other way to do it. Also, observe that the degree does not give a sharp count of the exact number of zeros; as in Bolzano's theorem, a change of sign over an interval guarantees that the function will vanishes, but we do not know how many times.

A closer look at the case $n = 1$ will give us a good clue as to how to deal with the general situation. Assume now that $f : [a,b] \to \mathbb{R}$ is a C^1 function and, for example, $f(a) < 0 < f(b)$. Although the number of zeros is not known, one may expect, at first glance, that this number is odd. Of course, this is not true since some awful things might happen:

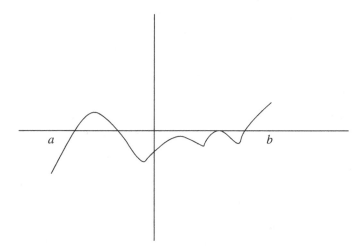

But there is an easy way to prevent such sad situations from happening: we can just assume that all the zeros of f are *simple*, that is, if f vanishes at x_0, then $f'(x_0) \neq 0$. This is enough to ensure that all zeros will be isolated and, furthermore, that the function will change sign at each zero. In this case, the degree can be computed just by adding the signs of the derivative over the set of zeros:

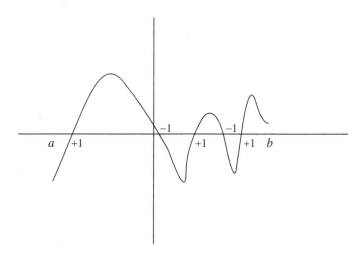

$$deg(f,\Omega,0) = \sum_{x \in f^{-1}(0)} sgn(f'(x)). \tag{5.7}$$

Observe, furthermore, that the last formula is valid for an arbitrary bounded open set Ω: because $f \neq 0$ on $\partial\Omega$, it is seen that f vanishes only in finitely many connected components of Ω; moreover, because all zeros are simple, the sum over each interval has only a finite number of terms.

It is clear that an arbitrary continuous f does not necessarily behave as nicely as the function in the last figure, but it can always be approximated by such a function: this is due to the Stone–Weierstrass theorem (which ensures the existence of smooth functions arbitrarily close to f) and to the *Sard lemma* (see Theorem 5.1 below). Adapted to this situation, the latter result implies that for any smooth function we may add an arbitrarily small constant in order to obtain a function g with the property that all its zeros are simple. In other words, this means that zero is a *regular value* of g, as follows from the following more general definition.

Definition 5.1. Let $U \subset \mathbb{R}^n$ be open and $f : U \to \mathbb{R}^m$ be of class C^1. A vector $p \in \mathbb{R}^m$ is called a regular value of f if $Df(x) : \mathbb{R}^n \to \mathbb{R}^m$ is surjective for all $x \in f^{-1}(p)$.

From this definition it is tautologically verified that all values $p \notin f(U)$ are regular. The values that are not regular are usually called *critical*; if $m > n$, then all values in $p \in f(U)$ are critical. In this last case, it is intuitively obvious that the set of critical values is small, but the result is also true for the following general case.

Theorem 5.1. *(Sard Lemma) Let $f : U \subset \mathbb{R}^n \to \mathbb{R}^m$ be a C^1 function. Then the set of critical values has zero measure. In particular, the set of regular values is dense in $\mathbb{R}^m$.*

The proof may be found in many textbooks (e.g., [87] and the original work [103]). In the present work, we shall only use this result for $m = n$ and f of class C^2. The particular case $m = n = 1$ is left as a simple and nice exercise for the reader; a short proof for arbitrary $m = n$ is given in the appendix of the present chapter.

Taking into account the previous comments on the case $n = 1$, when p is a regular value of f, it seems judicious to define $deg(f,\Omega,p)$ based on (5.7). With this idea in mind, let us firstly observe that if $f : \overline{\Omega} \to \mathbb{R}^n$ is a C^1 function such that $f \neq p$ on $\partial\Omega$ and p is a regular value of f, then the set $f^{-1}(p)$ is finite. Indeed, by the inverse function theorem, for each $x \in f^{-1}(p)$ there exist an open neighborhood U_x of x and an open neighborhood V_x of p such that $f : U_x \to V_x$ is a diffeomorphism. In particular, U_x does not contain any other preimage of p, and the claim follows from the compactness of $f^{-1}(p)$ and the fact that $f - p$ has no zeros on $\partial\Omega$. This allows the following definition, as seen in the following lemma.

Definition 5.2. Let $\Omega \subset \mathbb{R}^n$ be open and bounded, let $f : \overline{\Omega} \to \mathbb{R}^n$ be a C^1 mapping, and let $p \notin f(\partial\Omega)$ be a regular value of f. We define the Brouwer degree of f at p over Ω by

$$deg(f,\Omega,p) := \sum_{x \in f^{-1}(p)} sgn(Jf(x)), \tag{5.8}$$

where $Jf(x) := det(Df(x))$ denotes the Jacobian determinant of f at x.

Our next goal consists in extending the previous definition for arbitrary continuous f in such a way that (5.3)–(5.6) are satisfied. Unlike the case $n = 1$, this is not straightforward: as mentioned, the strategy will consist in choosing a smooth approximation of f that has p as a regular value, but some effort shall be needed in order to guarantee that the definition will not depend on this choice.

Before proceeding with the construction, let us show that if we want the four basic properties to be satisfied, then (5.8) is the only possible definition for the "smooth" case. After the construction is made, it will also become clear that the extension to continuous functions is also unique.

*Uniqueness

By (5.4) and (5.5), we may assume that $p = 0$ and that $f^{-1}(0)$ is a singleton. For simplicity, we may also assume, without loss of generality, that $f^{-1}(0) = \{0\}$ and, again by (5.5), that $\Omega = B_r(0)$ for arbitrarily small $r > 0$.

Using the fact that $Df(0)$ is invertible, we may fix a positive $\varepsilon < \inf_{v \in S^{n-1}} |Df(0)v|$ and, next, a value of r such that

$$f(x) = Df(0)x + R(x),$$

with $|R(x)| \leq \varepsilon |x|$ for $|x| < r$. We claim that f is homotopic to $g(x) := Df(0)x$. Indeed, for $x \neq 0$,

$$|Df(0)x + \lambda R(x)| = |x| \left| Df(0)\frac{x}{|x|} + \lambda \frac{R(x)}{|x|} \right|$$

$$\geq |x| \left(\inf_{v \in S^{n-1}} |Df(0)v| - \varepsilon \right) > 0.$$

Thus, it suffices to verify that $deg(g, B_r(0), 0) = sgn(Jf(0))$. Using homotopies of the type

$$h(x, \lambda) = A_{ij}(\lambda)Df(0)x \qquad \text{or} \qquad h(x, \lambda) = Df(0)A_{ij}(\lambda)x$$

for $i \neq j$, where the matrix $A_{ij}(\lambda)$ has a 1 at each entry over the diagonal, λc for some constant c at the ij entry, and zeros elsewhere, $Df(0)$ is transformed into a diagonal matrix. Furthermore, after replacing each of its entries a_{ii} (which, we recall, are nonzero) by $(1 - \lambda)a_{ii} + \lambda sgn(a_{ii})$, all the entries in the diagonal are transformed into ± 1. Next, if $a_{ii} = a_{jj} = -1$ for some $i \neq j$, then we define the homotopy

$$h(x, \lambda) = B_{ij}(\lambda)x,$$

where

$$B_{ij}(\lambda) = \begin{pmatrix} a_{11} & 0 & \cdots & \cdots & 0 \\ \cdots & \cdots & \cdots & \cdots & \cdots \\ \cdots & \cos((1+\lambda)\pi) & \cdots & \sin((1+\lambda)\pi) & 0 \\ \cdots & \cdots & \cdots & \cdots & \cdots \\ \cdots & \sin((1+\lambda)\pi) & \cdots & \cos((1+\lambda)\pi) & 0 \\ \cdots & \cdots & \cdots & \cdots & \cdots \\ 0 & 0 & \cdots & \cdots & a_{nn} \end{pmatrix}$$

which converts the -1 at the ii and jj entries into a 1. Thus, all possible situations have been reduced just to the following two cases:

- $Df(0) = Id$; here, $deg(g, \Omega, 0) = 1 = sgn(Jf(0))$;
- $Df(0)$ is a diagonal matrix with $a_{ii} = -1$ for some i and $a_{jj} = 1$ for all $j \neq i$. Here, $g(x) = (x_1, \ldots, -x_i, \ldots, x_n)$. Inspired by the case $n = 1$, we may consider for r small enough a cube C_1 of side r centered at 0 and an adjacent cube C_2 of side r centered at $(0, \ldots, 2r, \ldots, 0)$. Define $\tilde{g} : \overline{C_1 \cup C_2} \to \mathbb{R}^n$ by

$$\tilde{g}(x) = \begin{cases} g(x) & x \in \overline{C_1}, \\ (x_1, \ldots, x_i - 2r, \ldots, x_n) & x \notin \overline{C_1}. \end{cases}$$

Note that $\tilde{g}$ is homotopic to $(x_1, \ldots, r, \ldots, x_n)$, which does not vanish in $C_1 \cup C_2$; thus,

$$0 = deg(\tilde{g}, C_1 \cup C_2, 0) = deg(g, C_1, 0) + deg(\tilde{g}, C_2, 0).$$

The latter term is clearly equal to 1, so we conclude that

$$deg(f, \Omega, 0) = deg(g, C_1, 0) = -1 = sgn(Jf(0)),$$

and the proof is now complete.

Remark 5.1. As a consequence of the preceding analysis, it is seen that the general linear group $GL_n(\mathbb{R}) \subset \mathbb{R}^{n \times n}$ of invertible matrices has exactly two (pathwise) connected components. Indeed, we have shown that any invertible linear operator $T : \mathbb{R}^n \to \mathbb{R}^n$ is either homotopic to Id or to a diagonal matrix with $a_{ii} = -1$ for some i and $a_{jj} = 1$ for all $j \neq i$. Furthermore, it is easily seen that all the last cases are homotopic to one another. For example, observe that the homotopy

$$\begin{pmatrix} 2\lambda - 1 & \lambda(1 - \lambda) \\ \lambda(1 - \lambda) & 1 - 2\lambda \end{pmatrix}$$

transforms the matrix $\begin{pmatrix} -1 & 0 \\ 0 & 1 \end{pmatrix}$ into $\begin{pmatrix} 1 & 0 \\ 0 & -1 \end{pmatrix}$.

Existence

In this section, we shall prove that it is indeed possible to extend the previous definition to a (unique) function $deg : \{(f,\Omega,p) : f \in C(\overline{\Omega},\mathbb{R}^n), p \notin f(\partial\Omega)\} \rightarrow \mathbb{Z}$ satisfying (5.3)–(5.6). Although this is not, formally, a "starred" section, the readers who are not interested in technical details could skip it without a great damage. The only fundamental aspect to keep in mind is that the degree given by (5.8) in the regular case can indeed be extended for the general case and satisfies the properties described earlier. The brave ones may simply proceed with the construction.

To begin, we shall prove that if f is smooth and $p \notin f(\partial\Omega)$ is a regular value of f, then the degree defined by (5.8) can be expressed as an integral.

Lemma 5.1. *Let* $\Phi : \mathbb{R}^n \rightarrow \mathbb{R}^n$ *be a continuous function whose support is contained in* $B_\varepsilon(0)$ *for some small enough* ε *and such that* $\int_{\mathbb{R}^n} \Phi(x)\,dx = 1$. *Then*

$$deg(f,\Omega,p) = \int_\Omega \Phi(f(x) - p)Jf(x)\,dx. \tag{5.9}$$

Proof. The result is a consequence of the change of variables theorem. Indeed, take $\varepsilon > 0$ small enough and disjoint neighborhoods V_x of each preimage x of p such that $f|_{V_x} : V_x \rightarrow B_\varepsilon(p)$ is a diffeomorphism; then

$$\int_\Omega \Phi(f(y) - p)Jf(y)\,dy = \sum_{x \in f^{-1}(p)} \int_{V_x} \Phi(f(y) - p)Jf(y)\,dy$$

$$= \sum_{x \in f^{-1}(p)} sgn(Jf(x)) \int_{V_x} \Phi(f(y) - p)|Jf(y)|\,dy$$

$$= \sum_{x \in f^{-1}(p)} sgn(Jf(x)) \int_{B_\varepsilon(0)} \Phi(w)\,dw = deg(f,\Omega,p).$$

$\square$

Our next goal is to define the degree by formula (5.9) for *any* smooth mapping f such that $p \notin f(\partial\Omega)$. To this end, henceforth we shall take Φ as before, but also radial and vanishing in a neighborhood of 0, that is, $\Phi(x) = \phi(|x|)$, with $\phi : [0,+\infty) \rightarrow \mathbb{R}$ continuous such that

$$supp(\phi) \subset (0,\varepsilon), \qquad \int_{\mathbb{R}^n} \phi(|x|)\,dx = 1.$$

We claim that if $\varepsilon \leq dist(p, f(\partial\Omega))$, then the integral $\int_\Omega \phi(|f(x) - p|)Jf(x)\,dx$ does not depend on the choice of ϕ.

Lemma 5.2. *Let* $f : \overline{\Omega} \rightarrow \mathbb{R}^n$ *be a* C^1 *function with* $p \notin f(\partial\Omega)$, $\varepsilon \leq dist(p, f(\partial\Omega))$, *and* $\phi_j \in C([0,+\infty), \mathbb{R})$ *with support in* $(0,\varepsilon)$ *such that* $\int_{\mathbb{R}^n} \phi_j(|x|)\,dx = 1$ *for* $j = 1,2$. *Then* $\int_\Omega \phi_1(|f(x) - p|)Jf(x)\,dx = \int_\Omega \phi_2(|f(x) - p|)Jf(x)\,dx$.

Proof. By the Stone–Weierstrass theorem, we may assume that f is a C^2 function. Moreover, as $\phi(|f(x) - p|) = 0$ for x in a neighborhood of $\partial\Omega$, we may also assume that $\partial\Omega$ is smooth. Finally, we may assume that $p = 0$.

Set $\xi(s) = \phi_1(s) - \phi_2(s)$ and

$$\psi(s) := \frac{1}{s^n} \int_{B_s(0)} \xi(|x|)\,dx.$$

It is clear that ψ is of class C^1 and $supp(\psi) \subset (0, \varepsilon)$. Moreover,

$$s\psi'(s) = -n\psi(s) + s^{1-n}\frac{\partial}{\partial s}\left(\int_{B_s(0)} \xi(|x|)\,dx\right)$$

$$= -n\psi(s) + s^{1-n}\frac{\partial}{\partial s}\left(\omega_n \int_0^s \rho^{n-1}\xi(\rho)\,d\rho\right) = -n\psi(s) + \omega_n\xi(s),$$

where ω_n denotes the area of the unit sphere $S^{n-1} \subset \mathbb{R}^n$. On the other hand, define

$$V_j(x) := \psi(|f(x)|)detA_j(x),$$

where $A_j(x)$ is obtained by replacing the jth column of the Jacobian matrix of f at x by the vector $f(x)$. Although the norm function is not differentiable at 0, we took the precaution of taking $\psi \equiv 0$ in a neighborhood of 0, so the field $V = (V_1, \ldots, V_n)$ is of class C^1 and vanishes in a neighborhood of $\partial\Omega$. For convenience, denote by $M^{ij}(x)$ the Jacobian matrix of f at x, but with the ith row and the jth column replaced by zeros except for a 1 at the entry ij. Then

$$\sum_{j=1}^n \frac{\partial}{\partial x_j}[\psi(|f(x)|)]detA_j(x) = \sum_{j=1}^n \left(\frac{\psi'(|f(x)|)}{|f(x)|}\sum_{k=1}^n f_k(x)\frac{\partial f_k}{\partial x_j}(x)\right)detA_j(x)$$

$$= \frac{\psi'(|f(x)|)}{|f(x)|}\sum_{j,k=1}^n f_k(x)\frac{\partial f_k}{\partial x_j}(x)\sum_{i=1}^n(-1)^{i+j}f_i(x)det(M^{ij}(x)).$$

Rearranging terms and using the fact that

$$\sum_{j=1}^n(-1)^{i+j}\frac{\partial f_k}{\partial x_j}(x)det(M^{ij}(x)) = \begin{cases} Jf(x) & \text{if } k = i \\ 0 & \text{if } k \neq i \end{cases}.$$

we obtain

$$\sum_{j=1}^n \frac{\partial}{\partial x_j}[\psi(|f(x)|)]detA_j = \frac{\psi'(|f(x)|)}{|f(x)|}\sum_{i=1}^n f_i(x)^2 Jf(x) = Jf(x)|f(x)|\psi'(|f(x)|).$$

Moreover, regarding the determinant as a function from $\mathbb{R}^{n\times n}$ to $\mathbb{R}$ and expanding by the rth row or by the sth column, it is clear that $\frac{\partial det}{\partial a_{rs}}(A) = (-1)^{r+s}detA^{rs}$, where A^{rs} denotes the matrix obtained from A by deleting its rth row and its sth column. This implies

$$\frac{\partial}{\partial x_j}[detM^{ij}(x)] = \sum_{r,s=1}^{n} (-1)^{r+s} det([M^{ij}]^{rs}) \frac{\partial [M^{ij}]_{rs}}{\partial x_j}.$$

Clearly, $\frac{\partial [M^{ij}]_{rs}}{\partial x_j}$ is equal to $\frac{\partial^2 f^r}{\partial x_j \partial x_s}$ for $r \neq i, s \neq j$ and 0 in the remaining cases, so

$$\sum_{j=1}^{n} \frac{\partial}{\partial x_j}[detM^{ij}(x)] = \sum_{j=1}^{n} \sum_{r \neq i, s \neq j} (-1)^{r+s} det([M^{ij}]^{rs}) \frac{\partial^2 f^r}{\partial x_j \partial x_s}$$

$$= \sum_{s=1}^{n} \sum_{r \neq i, j \neq s} (-1)^{r+s} det([M^{ij}]^{rs}) \frac{\partial^2 f^r}{\partial x_s \partial x_j}.$$

Also, observe that, for $s \neq j$, $det([M^{ij}]^{rs}) = -det([M^{is}]^{rj})$, so we deduce that

$$\sum_{j=1}^{n} \frac{\partial}{\partial x_j}[detM^{ij}(x)] = -\sum_{s=1}^{n} \frac{\partial}{\partial x_s}[detM^{is}(x)].$$

Hence $\sum_{j=1}^{n} \frac{\partial}{\partial x_j}[detM^{ij}(x)] = 0$ and

$$\sum_{j=1}^{n} \frac{\partial}{\partial x_j}\left[\sum_{i=1}^{n} f_i(x) detM^{ij}(x)\right] = \sum_{i=1}^{n}\sum_{j=1}^{n} \frac{\partial f_i}{\partial x_j}(x) detM^{ij}(x) = nJf(x).$$

Summarizing, we obtain

$$divV(x) = Jf(x)\left[|f(x)|\psi'(|f(x)|) + n\psi(|f(x)|)\right] = \omega_n \xi(|f(x)|)Jf(x),$$

and thus, by the Gauss divergence theorem,

$$\int_\Omega \xi(|f(x)|)Jf(x)\,dx = \frac{1}{\omega_n}\int_{\partial\Omega} divV(x)\,dx = \frac{1}{\omega_n}\int_{\partial\Omega} V(x)\cdot d\mathbf{S} = 0,$$

which proves that $\int_\Omega \phi_1(|f(x)|)Jf(x)\,dx = \int_\Omega \phi_2(|f(x)|)Jf(x)\,dx$. $\square$

We are now in a position to prove the following lemma, which will allow us to define the degree for arbitrary smooth functions and, later on, for continuous functions.

Lemma 5.3. *Let $f \in C^1(\overline{\Omega}, \mathbb{R}^n)$, and let p be a regular value of f such that $r := dist(p, f(\partial\Omega)) > 0$. If $g \in C^1(\overline{\Omega}, \mathbb{R}^n)$ verifies $\|f - g\|_\infty < \frac{r}{7}$ and p is a regular value of g, then $deg(g, \Omega, p) = deg(f, \Omega, p)$.*

Proof. As before, we may assume $p = 0$. Set $\varepsilon := \frac{r}{7}$ and consider a C^1 mapping $\eta : [0, +\infty) \to [0, 1]$ such that $\eta \equiv 1$ on $[0, 2\varepsilon]$ and $\eta \equiv 0$ on $[3\varepsilon, +\infty)$. Then the function

$$h(x) := (1 - \eta(|f(x)|))f(x) + \eta(|f(x)|)g(x)$$

is of class C^1, with

$$h(x) = f(x) \qquad \text{if } |f(x)| < 2\varepsilon,$$

$$h(x) = g(x) \qquad \text{if } |f(x)| > 3\varepsilon.$$

Moreover,

$$\|h - f\|_\infty < \varepsilon, \quad \|h - g\|_\infty < \varepsilon$$

and

$$|g(x)| > dist(0, f(\partial\Omega)) - \varepsilon, \qquad |h(x)| > dist(0, f(\partial\Omega)) - \varepsilon$$

for all $x \in \partial\Omega$. Next, take a function $\phi \in C(\mathbb{R}, \mathbb{R})$ with support in $(0, \varepsilon)$ such that $\int_{\mathbb{R}^n} \phi(|x|)\, dx = 1$, and let $\tilde{\phi}(s) = \phi(s - 4)$. Observe that

$$supp(\phi), supp(\tilde{\phi}) \subset (0, 5\varepsilon), \quad 5\varepsilon < dist(0, h(\partial\Omega)),$$

so we obtain

$$\int_\Omega \phi(|h(x)|)Jh(x)\, dx = \int_\Omega \tilde{\phi}(|h(x)|)Jh(x)\, dx.$$

On the other hand, $\phi(y) = 0$ for $|y| \geq \varepsilon$ and $|f(x)| < 2\varepsilon$ when $|h(x)| < \varepsilon$; thus,

$$\phi(|h(x)|)Jh(x) = \phi(|f(x)|)Jf(x)$$

for all x. In the same way, $\tilde{\phi}(y) = 0$ for $|y| \leq 4\varepsilon$, and if $|h(x)| > 4\varepsilon$, then $|f(x)| > 3\varepsilon$. Hence

$$\tilde{\phi}(|h(x)|)Jh(x) = \tilde{\phi}(|g(x)|)Jg(x)$$

for all x. This proves that

$$\int_\Omega \phi(|f(x)|)Jf(x)\, dx = \int_\Omega \tilde{\phi}(|g(x)|)Jg(x)\, dx$$

and so completes the proof. $\square$

Recall that, from Theorem 5.1, the set RV of regular values of a function f is dense, so by the preceding lemma we may define the degree for arbitrary smooth functions. Our proof of Sard's lemma in the appendix requires more regularity, so we shall assume for the moment that the mapping is C^2.

Definition 5.3. Let $f \in C^2(\overline{\Omega}, \mathbb{R}^n)$ such that $p \notin f(\partial\Omega)$, and define

$$deg(f, \Omega, p) := \lim_{q \to p, q \in RV} deg(f, \Omega, q).$$

To verify that the degree is well defined, let us consider, for example, $\varepsilon < \frac{1}{15} dist$ $(p, f(\partial\Omega))$. If q_1 and q_2 are regular values of f such that $|q_j - p| < \varepsilon$ for $j = 1, 2$, then set $g_j := f - q_j + p$. Note that p is a regular value of both functions and, moreover,

$$\|g_j - f\|_\infty < \varepsilon, \quad dist(p, g_j(\partial\Omega)) \geq dist(p, f(\partial\Omega)) - \varepsilon > 14\varepsilon.$$

Since $\|g_1 - g_2\|_\infty < 2\varepsilon$, the previous lemma implies that $deg(g_1, \Omega, p) = deg(g_2, \Omega, p)$.

To proceed with the last step, let us firstly observe that if f is continuous and $p \notin f(\partial\Omega)$, then $p \notin g(\partial\Omega)$ when g is close enough to f for the uniform norm.

Definition 5.4. Let $f \in C(\overline{\Omega}, \mathbb{R}^n)$ such that $p \notin f(\partial\Omega)$. We define

$$deg(f, \Omega, p) := \lim_{\|g-f\|_\infty \to 0, g \in C^2} deg(g, \Omega, p).$$

Besides the Stone–Weierstrass theorem, once again we deduce that the degree is well defined from the fact that, if two C^2 functions are close to one another for the uniform norm and p is not at the image of $\partial\Omega$, then both of them have the same degree. An elegant way to show this consists in defining the set

$$\mathscr{C}_p := \{f \in C(\overline{\Omega}, \mathbb{R}^n) : p \notin f(\partial\Omega)\}$$

and noticing that the mapping $deg(\cdot, \Omega, p)$ is locally constant in $\mathscr{C}_p \cap C^2(\overline{\Omega}, \mathbb{R}^n)$. The connected components of this latter set coincide with those of $\mathscr{C}_p$ intersected with $C^2(\overline{\Omega}, \mathbb{R}^n)$, so there exists a unique extension of the degree to $\mathscr{C}_p$, given precisely by the preceding definition.

To conclude, let us check that properties (5.3)–(5.6) are verified. The first one is obvious since p is a regular value of the identity function. The other three properties are deduced from the fact that the degree is constant over the connected components of $\mathscr{C}_p$. Indeed, (5.4) and (5.5) hold when f is C^2 and p is a regular value, so they hold for any continuous f since (by Sard's lemma and the Stone–Weierstrass theorem) each connected component of $\mathscr{C}_p$ contains a smooth function having p as a regular value.

Finally, the homotopy invariance (5.6) can be proven in the following way. Assume that $h : \overline{\Omega} \times [0, 1] \to \mathbb{R}^n$ is continuous with $h(x, \lambda) \neq p$ for $x \in \partial\Omega$ and $\lambda \in [0, 1]$ such that $h(x, 0) = f(x)$ and $h(x, 1) = g(x)$. The curve $\varphi : [0, 1] \to \mathscr{C}_p$ given by $\varphi(\lambda) := h(\cdot, \lambda)$ is continuous, so all the functions $h(\cdot, \lambda)$ belong to the same connected component. Furthermore, the same argument shows that it is also possible to move the point p continuously, as can be seen in the following exercise.

Exercise 5.1. Let $h : \overline{\Omega} \times [0, 1] \to \mathbb{R}^n$ and $p : [0, 1] \to \mathbb{R}^n$ be continuous such that $h(x, \lambda) \neq p(\lambda)$ for $x \in \partial\Omega$ and $\lambda \in [0, 1]$. Prove that $deg(h(\cdot, \lambda), \Omega, p(\lambda))$ is constant.

It is worth mentioning that the degree is also invariant under smooth changes of the domain. In more precise terms, if $\Omega \subset \mathbb{R}^n \times [0, 1]$ is open and bounded and $\Omega_\lambda := \{x \in \mathbb{R}^n : (x, \lambda) \in \Omega\}$, then for any $h : \overline{\Omega} \to \mathbb{R}^n$ continuous and $p \in \mathbb{R}^n$ such that $p \notin h(\partial\Omega)$ the degree $deg(h(\cdot, \lambda), \Omega_\lambda, p)$ is well defined and does not depend on λ. For a proof, see, for example, [30]; we may sketch an alternative proof as follows.

Assume $p = 0$, and set $h_\lambda = h(\cdot, \lambda)$. For fixed λ_0 and $y = (x, \lambda_0) \in \partial\Omega$, there exist $r_y, \varepsilon_y > 0$ such that h does not vanish in the closure of $B_{r_y}(x) \times (\lambda_0 - \varepsilon_y, \lambda_0 + \varepsilon_y)$. The compact set $(\partial\Omega)_{\lambda_0}$ is covered by finitely many of these sets, so

$$(\partial\Omega)_{\lambda_0} \subset \bigcup_{j=1}^{N} B_{r_j}(x_j).$$

Now define the open set $\Omega_0 := \Omega_{\lambda_0} \setminus \bigcup_{j=1}^{N} \overline{B_{r_j}(x_j)}$. We claim that if $\delta > 0$ is small enough, then $\Omega_0 \subset \Omega_\lambda$ for $\lambda \in (\lambda_0 - \delta, \lambda_0 + \delta)$ and h_λ does not vanish outside Ω_0.

Indeed, on the one hand, suppose that there exist $\lambda_n \to \lambda_0$ and $x_n \in \Omega_0 \setminus \Omega_{\lambda_n}$; then, passing to a subsequence, we may suppose that $x_n \to x_0 \in \overline{\Omega_0}$. Thus, $(x_0, \lambda_0) \in \overline{\Omega}$ and, furthermore, $(x_n, \lambda_n) \notin \Omega$ so $(x_0, \lambda_0) \in \partial\Omega$. We deduce that $x_0 \in B_{r_j}(x_j)$ for some j and $x_n \in B_{r_j}(x_j)$ for n large, a contradiction since $x_n \in \Omega_0$. On the other hand, if $h(x_n, \lambda_n) = 0$ for some $\lambda_n \to \lambda_0$ and $x_n \notin \Omega_0$, then, taking a subsequence, we may assume $x_n \to x_0 \notin \Omega_0$. But $h(x_0, \lambda_0) = 0$, and hence $x_0 \notin \bigcup_{j=1}^{N} \overline{B_{r_j}(x_j)}$, a contradiction. Hence, for λ close enough to λ_0 we obtain, by excision and homotopy invariance,

$$deg(h_\lambda, \Omega_\lambda, 0) = deg(h_\lambda, \Omega_0, 0) = deg(h_{\lambda_0}, \Omega_0, 0) = deg(h_{\lambda_0}, \Omega_{\lambda_0}, 0).$$

This proves that $deg(h_\lambda, \Omega_\lambda, 0)$ is locally constant and, thus, constant over $[0, 1]$.

Remark 5.2. The property is not true under the weaker assumption $p \notin h_\lambda(\partial\Omega_\lambda)$ for all λ. This might look surprising at first sight, but the reason is easily understood from this example. Consider

$$\Omega = \Omega_1 \cup \Omega_2 \subset \mathbb{R} \times [0, 1],$$

where

$$\Omega_1 = \left\{ (x, \lambda) : 2\lambda - 1 < x < 1 - 2\lambda, 0 \le \lambda < \frac{1}{2} \right\},$$

$$\Omega_2 = \left\{ (x, \lambda) : 1 - 2\lambda < x < 2\lambda - 1, \frac{1}{2} < \lambda \le 1 \right\},$$

and let

$$h(x, \lambda) = \begin{cases} x & \text{if } \lambda \le \frac{1}{2}, \\ |x| & \text{if } \lambda > \frac{1}{2}. \end{cases}$$

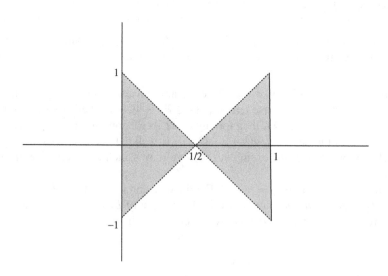

Then $h : \overline{\Omega} \to \mathbb{R}$ is continuous and

- If $\lambda < \frac{1}{2}$, then $\Omega_\lambda = (2\lambda - 1, 1 - 2\lambda)$ and $h_\lambda = Id$, so $deg(h_\lambda, \Omega_\lambda, 0) = 1$;
- If $\lambda > \frac{1}{2}$, then $\Omega_\lambda = (1 - 2\lambda, 2\lambda - 1)$ and $h_\lambda|_{\partial\Omega_\lambda} > 0$, so $deg(h_\lambda, \Omega_\lambda, 0) = 0$.

Note that $\Omega_{1/2} = \emptyset$, so clearly $0 \notin h_{1/2}(\partial\Omega_{1/2})$.

5.1.1 Comments, Examples, and Applications

The construction of the preceding section shows that the degree is defined in a unique way for continuous functions $f : \overline{\Omega} \to \mathbb{R}^n$ such that $p \notin f(\partial\Omega)$. In particular, when $n = 2$ and f, regarded as a complex function, is analytic, we get a better understanding of why formula (5.1) counts so precisely the exact number of zeros. Indeed, the *Cauchy–Riemann* conditions imply that the Jacobian matrix of f at a given z has the form

$$\begin{pmatrix} a & -b \\ b & a \end{pmatrix}$$

where a and b are respectively the real and imaginary parts of $f'(z)$. Note that the Jacobian determinant $a^2 + b^2$ is nonnegative, and zero if and only if $f'(z) = 0$; in particular, if all the zeros of f are simple, then all signs in the sum (5.8) are positive. When f has multiple zeros, a well-known result in complex analysis says that we may add a small complex number so the resulting function has only simple zeros; this is, again, a particular case of Sard's lemma. A trivial example is $f(z) = z^k$, whose degree over a ball $B_r(0)$ can be easily computed by formula (5.1), but even more easily by adding a small constant: if $0 < |w| < \sqrt[k]{r}$, then $g(z) := z^k - w$ has exactly k simple roots in $B_r(0)$, so its degree is k. To complete the picture, we may recall the aforementioned result: if an analytic function has a zero at z_0 with multiplicity k, then there exist neighborhoods U and V of z_0 and 0 such that the equation $f(z) = w$ has exactly k different simple solutions in U for each $w \in V \setminus \{0\}$. Of course, we cannot pretend to obtain such a good result when f is just continuous, but now we are able to give a more precise account of what we meant by a "robust" set of zeros at the beginning of the chapter. If $deg(f, \Omega, 0) \neq 0$, then f vanishes in Ω, but we are saying much more: among other things, that the equation $g(x) = 0$ has at least one solution in Ω for g close enough to f. Taking $g = f + c$ with c constant, we deduce as a consequence that $f(\Omega)$ contains a neighborhood of 0. This fact can be regarded as a generalization of the open mapping theorem, which establishes that a nonconstant analytic function over a connected open subset of $\mathbb{C}$ is open.

Also, Rouché's theorem admits a convenient extension to the present context: if $f, g : \overline{\Omega} \to \mathbb{R}^n$ are continuous and

$$|f(x) - g(x)| < |f(x)| \qquad \text{for all} \quad x \in \partial\Omega,$$

then $deg(g, \Omega, 0) = deg(f, \Omega, 0)$. Indeed, observe in the first place that f and g cannot vanish on $\partial\Omega$, so both degrees are well defined; moreover, the linear homotopy

$$h(x, \lambda) = (1 - \lambda)f(x) + \lambda g(x)$$

never vanishes for $x \in \partial\Omega$ since

$$|h(x, \lambda)| = |f(x) - \lambda(f(x) - g(x))| \geq |f(x)| - |f(x)) - g(x)| > 0.$$

This result is also called the *dog-on-a-leash* theorem and has an obvious interpretation when $n = 2$: if a person walks his/her dog and the length of the leash is possibly variable but always shorter than the distance from the dog's owner to the origin, then both of them (dog and owner) perform the same number of turns around 0.

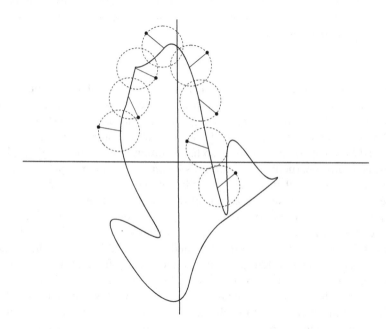

There is a more general version, usually attributed to Easterman, which only assumes that $f \neq 0$ over $\partial\Omega$ and

$$|f(x) - g(x)| < |f(x)| + |g(x)| \qquad \text{for all} \quad x \in \partial\Omega.$$

In other words, this weaker condition allows the leash to be as long as the dog wants, as long as the origin is never situated in the segment between the dog and the owner (some more imaginative interpretations assume the existence of a tree at $x = 0$, so the assumption says that the tree never obscures the view of the dog).

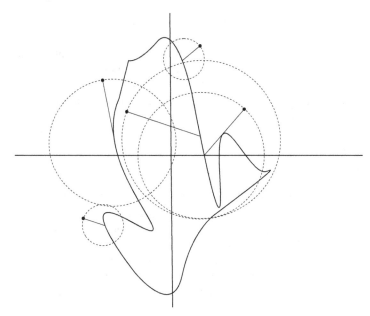

The proof employs the same homotopy but a slightly sharper estimate: from the inequalities

$$|h(x,\lambda)| = |f(x) - \lambda(f(x) - g(x))| \geq |f(x)| - \lambda|f(x) - g(x)|$$

and

$$|h(x,\lambda)| = |(1-\lambda)(f(x) - g(x)) + g(x)| \geq |g(x)| - (1-\lambda)|f(x) - g(x)|$$

it follows that

$$2|h(x,\lambda)| \geq |f(x)| + |g(x)| - |f(x) - g(x)| > 0$$

for $x \in \partial\Omega$, so the proof is complete.

A different and useful example is the case of a continuous function $f : \overline{B_R(0)} \to \mathbb{R}^n$ satisfying

$$f(x) \cdot x > 0 \qquad \text{for} \quad |x| = R.$$

Here, the degree computation is also immediate by the homotopy invariance: again, it suffices to define the linear homotopy

$$h(x,\lambda) = \lambda f(x) + (1-\lambda)x,$$

which does not vanish for $x \in \partial B_R(0)$ since $h(x,\lambda) \cdot x = \lambda f(x) \cdot x + (1-\lambda)R^2 > 0$. But $h(\cdot,0)$ is the identity function, so we deduce that

$$deg(f, B_R(0), 0) = deg(h(\cdot,1), B_R(0), 0) = deg(Id, B_R(0), 0) = 1.$$

In particular, this provides a very short proof of Theorem 4.3. The proof of the no-retraction theorem is even shorter: if $r : \overline{B_1(0)} \to \partial B_1(0)$ is continuous and $r = Id$ over $\partial B_1(0)$, then $deg(r, B_1(0), 0) = 1$, and hence it vanishes, a contradiction.

A similar argument can be used for a proof of the Poincaré–Miranda theorem. Also, Brouwer's theorem follows in a direct way: let $f : \overline{B_1(0)} \to \overline{B_1(0)}$ be continuous. If $f(x) = x$ for some $x \in \partial B_1(0)$, then there is nothing to prove; otherwise, define $h(x, \lambda) = x - \lambda f(x)$. If $h(x, \lambda) = 0$ for some $x \in \partial B_1(0)$, then $x = \lambda f(x)$ and $1 = |x| = \lambda |f(x)| \le \lambda$. This implies $\lambda = 1$ and $f(x) = x$, a contradiction. Thus $deg(f, B_1(0), 0) = deg(Id, B_1(0), 0) = 1$.

Another well-known theorem in algebraic topology that can be easily proven using degree theory is the *hairy ball theorem*, firstly stated by Poincaré in the nineteenth century and demonstrated later by Brouwer. Roughly speaking, the theorem says that any continuous vector field over an even-dimensional sphere S is orthogonal to S at some points; in particular, there are no continuous nonvanishing tangent vector fields. This explains its name, in the sense that "no one can comb a coconut." However, the latter does not seem to be a very practical application, so we may give another "real-life" interpretation, particularly useful for those who are planning to go on a picnic: at any given moment there exists at least one point on Earth with no wind.

Theorem 5.2. *Let $f : \overline{B_1(0)} \to \mathbb{R}^n$ be continuous, and assume that n is odd. Then there exist $r \in \mathbb{R}$ and $x \in \partial B_1(0)$ such that $f(x) = rx$. In particular, there are no nonvanishing continuous tangent fields over the sphere S^{n-1}.*

Proof. Suppose $f(x) \ne rx$ for all $r \in \mathbb{R}$ and $x \in \partial B_1(0)$, and define the homotopies

$$h_{\pm}(x, \lambda) = \lambda f(x) \pm (1 - \lambda)x.$$

If $h_{\pm}(x, \lambda) = 0$ for some $\lambda \in [0, 1]$ and $x \in \partial B_1(0)$, then $\lambda > 0$ and $f(x) = \pm \frac{\lambda - 1}{\lambda} x$, which contradicts the fact that $f(x) \ne rx$. Thus,

$$deg(f, B_1(0), 0) = deg(\pm Id, B_1(0), 0) = (\pm 1)^n,$$

a contradiction, since n is odd.

Moreover, if $T : S^{n-1} \to \mathbb{R}^n$ is continuous and $T(x) \cdot x = 0$ for all x, then, using Tietze's theorem (Thm. 5.9), we may extend T continuously to $\overline{B_1(0)}$ to deduce that $T(x) = rx$ for some $r \in \mathbb{R}$ and some $x \in S^{n-1}$. Thus, $0 = T(x) \cdot x = rx \cdot x = r|x|^2 = r$, and hence $T(x) = 0$. $\square$

Theorem 5.3. *(Borsuk) Let $\Omega \subset \mathbb{R}^n$ be a bounded open symmetric neighborhood of 0, and let $f : \overline{\Omega} \to \mathbb{R}^n$ be continuous and odd, that is, $f(-x) = -f(x)$ for all x, with $0 \notin f(\partial \Omega)$. Then $deg(f, \Omega, 0)$ is odd.*

Proof. Fix $\varepsilon > 0$ such that $\overline{B_\varepsilon(0)} \subset \Omega$, and define $V = \Omega \backslash \overline{B_\varepsilon(0)}$ y $\varphi : \partial V \to \mathbb{R}^n$ given by

$$\varphi(x) = \begin{cases} f(x) & x \in \partial \Omega, \\ x & |x| = \varepsilon. \end{cases}$$

By the extension of Tietze's theorem given in Lemma 5.8, we may suppose that φ is an odd continuous mapping defined on $\overline{V}$ with $\varphi(x) \neq 0$ when $x_n = 0$. Next, define $V^{\pm} = \{x \in V : \pm x_n > 0\}$, so

$$deg(\varphi, V, 0) = deg(\varphi, V^+, 0) + deg(\varphi, V^-, 0)$$

and, because φ is odd, it is seen that $deg(\varphi, V^+, 0) = deg(\varphi, V^-, 0)$. Indeed, it suffices to approximate φ on $\overline{V^+}$ by a C^1 function ψ and select a regular value z of ψ such that $|z|$ is small enough thus, $\psi^-(x) := -\psi(-x)$ is an approximation of φ on $\overline{V^-}$ and $-z$ is a regular value. Hence

$$deg(\varphi, V^-, 0) = deg(\psi^-, V^-, -z) = \sum_{\psi^-(x)=-z} sgn(J\psi^-(x))$$

$$= \sum_{\psi(x)=z} sgn(J\psi(x)) = deg(\psi, V^+, z) = deg(\varphi, V^+, 0).$$

Extending φ as the identity inside $B_\varepsilon(0)$, we conclude

$$deg(f, \Omega, 0) = deg(\varphi, \Omega, 0) = deg(\varphi, B_\varepsilon(0), 0) + deg(\varphi, V, 0) = 1 + 2deg(\varphi, V^+, 0),$$

and the proof is complete. $\square$

Alternative proof: Using Exercise 5.6, we may assume that f is a C^1 function and $Jf(0) \neq 0$. If 0 is a regular value of f, then the result is clear since

$$deg(f, \Omega, 0) = sgn(Jf(0)) + \sum_{x \in f^{-1}(0) \setminus \{0\}} sgn(Jf(x)),$$

and the sum has an even number of terms. Otherwise, we may "regularize" 0 as follows.

Let $\Omega_1 := \{x \in \Omega : x_1 \neq 0\}$, let y^1 be a regular value of the function $g(x) := \frac{f(x)}{x_1^3}$ in Ω_1, and define $f^1(x) := f(x) - x_1^3 y^1$. By Sard's lemma, y^1 can be chosen arbitrarily close to 0, so we may suppose $deg(f^1, \Omega, 0) = deg(f, \Omega, 0)$. On the other hand, if $x \in \Omega_1$ is a zero of f^1, then $g(x) = y^1$ and $Df^1(x) = x_1^3 Dg(x)$ since:

- If $j \neq 1$, then $\frac{\partial f^1}{\partial x_j} = \frac{\partial f}{\partial x_j} = x_1^3 \frac{\partial g}{\partial x_j}$;
- $\frac{\partial f^1}{\partial x_1} = \frac{\partial f}{\partial x_1} - 3x_1^2 y^1 = \frac{\partial f}{\partial x_1} - 3x_1^2 g(x) = x_1^3 \frac{\partial g}{\partial x_1}$.

We deduce that 0 is a regular value of f^1 in Ω_1. Inductively, we may choose a regular value y^{k+1} of $\frac{f^k}{x_{k+1}^3}$ in Ω_{k+1} close enough to 0 and define $f^{k+1}(x) = f^k(x) - x_{k+1}^3 y^{k+1}$. Then 0 is a regular value of f^k in Ω_k and $deg(f, \Omega, 0) = deg(f^k, \Omega, 0)$ for all k.

We claim that 0 is a regular value of f^n in Ω. Indeed, if $f^n(x) = 0$, then we already know that $Df(x)$ is invertible when $x \in \Omega_n$. If $x \notin \Omega_n$, then $x_n = 0$ and

$$f^n(x) = f^{n-1}(x), \qquad Df^n(x) = Df^{n-1}(x).$$

Iterating the procedure, if $x \neq 0$, then $f^n(x) = f^j(x)$ and $Df^n(x) = Df^j(x)$, where x_j is the last nonzero coordinate of x. Note that $x \in \Omega_j$, so we deduce that $Df^j(x)$ is invertible. Finally, for $x = 0$ we obtain $Df^n(0) = Df(0)$, which is also invertible.

Corollary 5.1. *(Borsuk–Ulam) Let $f : S^n \to \mathbb{R}^n$ be continuous. Then there exists x such that $f(x) = f(-x)$.*

Proof. Let $\varphi(x) := f(x) - f(-x)$. If φ does not vanish, extend it to $\overline{B_1(0)} \subset \mathbb{R}^{n+1}$ by

$$\psi(x) := \begin{cases} |x|\varphi\left(\frac{x}{|x|}\right) & \text{if } x \neq 0, \\ 0 & \text{if } x = 0. \end{cases}$$

Next, identify $\mathbb{R}^n$ with $\mathbb{R}^n \times \{0\} \subset \mathbb{R}^{n+1}$, and deduce from the Borsuk theorem that $deg(\psi, B, 0)$ is odd. On the other hand, we may consider $y = (0, \dots, 0, \varepsilon)$ with $\varepsilon > 0$ small enough such that $deg(\psi, B_1(0), 0) = deg(\psi, B_1(0), y)$. But $y \notin \varphi(B_1(0))$, so $deg(\psi, B_1(0), y) = 0$, a contradiction. □

There is a very popular application of the Borsuk–Ulam theorem for the specific case $n = 2$, again of a meteorological character: at any given moment there exists on Earth a pair of antipodal points with the same temperature and pressure.

In addition, the Borsuk–Ulam theorem provides a different proof of Brouwer's theorem. Suppose that $f : \overline{B_1(0)} \to \overline{B_1(0)}$ is continuous with no fixed points, and let $F : \overline{B_1(0)} \to \mathbb{R}^n$ be defined by $F(x) := (1 - |x|)f(x) + |x|x$. Then $F = Id$ over $\partial B_1(0)$, so the function $g : S^n \subset R^{n+1} \to \mathbb{R}^n$ given by

$$g(x_1, \dots, x_{n+1}) := \begin{cases} F(x_1, \dots, x_n) & \text{if } x_{n+1} > 0 \\ (x_1, \dots, x_n) & \text{if } x_{n+1} \leq 0 \end{cases}$$

is continuous and satisfies $g(x) \neq g(-x)$ for all $x \in S^n$, a contradiction.

For those who are not yet completely convinced, Borsuk's theorem gives two more reasons to believe that Brouwer's theorem (in fact, the no-retraction theorem) is true. First, observe that if Ω is a symmetric open bounded neighborhood of 0 and $f : \partial\Omega \to \mathbb{R}^n$ is continuous and odd, then any continuous extension of f to $\overline{\Omega}$ has least a zero and a fixed point. This is due to the fact that the degree of the extended function is odd, so it vanishes and $f(x) - x$ is also an odd continuous function, so it vanishes as well. Now, suppose that $r : \overline{B_1(0)} \to \partial B_1(0)$ is a continuous retraction. In particular, r is odd, and hence it vanishes, a contradiction. Moreover, $-r$ is also odd, so it has a fixed point and this is a new contradiction. All of this is obviously unnecessary since, as mentioned a few pages ago, the fact that r coincides on $\partial B_1(0)$ with the identity function implies directly that its degree is 1; nevertheless, these last remarks may give the reader a more general picture of the situation.

Another famous application of the Borsuk–Ulam theorem is the following Lusternik–Schnirelmann covering theorem.

Theorem 5.4. *Let $\{C_1, \dots, C_{n+1}\}$ be a closed covering of S^n. Then there exists j such that C_j contains antipodal points.*

Proof. Suppose that $A(C_j) \cap C_j = \emptyset$ for $j = 1, \ldots, n$, where $A : S^n \to S^n$ denotes the antipodal mapping $A(x) := -x$. Define the continuous mapping $u : S^n \to \mathbb{R}^n$ given by $u = (u_1, \ldots, u_n)$, where u_j is the Urysohn function

$$u_j(x) = \frac{dist(x, C_j)}{dist(x, C_j) + dist(x, A(C_j))},$$

which separates C_j and $A(C_j)$ (that is, $u_j|_{C_j} \equiv 0$ and $u_j|_{A(C_j)} \equiv 1$). From the Borsuk–Ulam theorem, there exists $x \in S^n$ such that $u(x) = u(A(x))$ and, hence, $x \notin C_j \cup A(C_j)$ for $j = 1, \ldots, n$. In other words, $x \notin \bigcup_{j=1}^n C_j$ and $-x \notin \bigcup_{j=1}^n C_j$, so C_{n+1} contains both x and $-x$. $\quad\square$

But perhaps one of the most relevant topological consequences of Borsuk's theorem is the result that, among other things, says that open subsets of $\mathbb{R}^n$ and $\mathbb{R}^m$ are not homeomorphic when $m \neq n$. The following theorem states this in a more general way.

Theorem 5.5. *(Invariance of domain) Let $U \subset \mathbb{R}^n$ be open, and let $f : U \to \mathbb{R}^n$ be locally injective, that is, for every $x \in U$ there exists $r > 0$ such that f is injective in $B_r(x)$. Then f is open. In particular, if $V \subset \mathbb{R}^m$ is open and homeomorphic to U, then $m = n$.*

Proof. It suffices to prove that if $f : \overline{B_r(0)} \to \mathbb{R}^n$ is an injective continuous function with $f(0) = 0$, then $f(B_r(0))$ is a neighborhood of 0. Let $h : \overline{B_r(0)} \times [0, 1] \to \mathbb{R}^n$ be given by

$$h(x, \lambda) = f\left(\frac{x}{1+\lambda}\right) - f\left(\frac{-\lambda x}{1+\lambda}\right);$$

then $h(x, 0) = f(x)$ and $h(x, 1) = f\left(\frac{x}{2}\right) - f\left(-\frac{x}{2}\right)$, which is an odd function. Moreover, if $h(x, \lambda) = 0$, then $f\left(\frac{x}{1+\lambda}\right) = f\left(\frac{-\lambda x}{1+\lambda}\right)$, and hence $x = 0 \notin \partial B_r(0)$. This proves that $deg(f, B_r(0), 0)$ is odd; in particular, $deg(f, B_r(0), p) \neq 0$ for every p close enough to 0.

Now assume that $\varphi : U \to V$ is a homeomorphism and suppose, for example, that $n > m$. Identifying $\mathbb{R}^m$ with $\mathbb{R}^m \times \{0\} \subset \mathbb{R}^n$, we conclude that $\varphi : U \to \mathbb{R}^n$ is open, a contradiction since $\varphi(U) \subset \mathbb{R}^m$. $\quad\square$

An immediate application is the following: if $f : \mathbb{R}^n \to \mathbb{R}^n$ is continuous, locally injective, and *coercive*, in the sense that $|f(x)| \to \infty$ for $|x| \to \infty$, then f is surjective. Indeed, local injectivity implies that the range of f is an open set; moreover, if $f(x_n) \to y$, then the coerciveness assumption implies that $\{x_n\}$ is bounded. Thus, there exists a subsequence that converges to some x and, by continuity, $f(x) = y$. This proves that $Im(f)$ is also closed and, hence, $Im(f) = \mathbb{R}^n$.

It is worth noting that the invariance of the domain implies that if $U \subset \mathbb{R}^n$ is open and $f : U \to \mathbb{R}^n$ is continuous and injective, then $f(U)$ is open and $f^{-1} : f(U) \to U$ is also continuous. Thus, it is no great surprise to discover that a version of the implicit function theorem for continuous functions can be obtained as well, as follows.

Theorem 5.6. *Let $U \subset \mathbb{R}^n$ and $V \subset \mathbb{R}^m$ be open, and let $f : U \times V \to \mathbb{R}^m$ be continuous with $f(x_0, y_0) = 0$ for some $(x_0, y_0) \in U \times V$. If $f(x, \cdot) : V \to \mathbb{R}^m$ is injective for all x, then there exist a $U_0 \subset U$ neighborhood of x_0 and a unique $\phi : U_0 \to V$ continuous such that $\phi(x_0) = y_0$ and $f(x, \phi(x)) = 0$ for all $x \in U_0$.*

Proof. Define $F : U \times V \to \mathbb{R}^{n+m}$ by $F(x, y) = (x, f(x, y))$. Clearly, F is continuous and injective, so it has a continuous inverse $F^{-1} : F(U \times V) \to U \times V$.

Now let $U_0 := \{x \in U : (x, 0) \in F(U \times V)\}$, and define $\phi : U_0 \to V$ as the second coordinate of $F^{-1}(x, 0)$. Then ϕ is continuous and $f(x, \phi(x)) = 0$. Uniqueness is obvious from the injectivity of $f(x, \cdot)$; in particular, $\phi(x_0) = y_0$. □

We end this section with an application to boundary value problems. Because the Brouwer degree is defined for finite-dimensional mappings, we cannot use it yet for abstract equations $Fu = 0$, although it may be of great help when applying shooting type methods. For example, let us consider the periodic problem

$$u'(t) + f(t, u(t)) = 0, \qquad u(0) = u(1), \tag{5.10}$$

with $f : [0, 1] \times \mathbb{R}^n \to \mathbb{R}^n$ continuous and locally Lipschitz in u. Suppose that $\Omega \subset \mathbb{R}^n$ is open and bounded with the property that all the solutions of the initial value problem starting at $\overline{\Omega}$ are defined up to $t = 1$. In this case, the Poincaré operator P is defined in $\overline{\Omega}$. Under appropriate assumptions, it is possible to prove that the degree of the mapping $F := Id - P$ is well defined and different from 0.

For example, let us define for $\lambda \in [0, 1]$ the function $P_\lambda(u_0) := u(\lambda)$, where u is the unique solution of the equation with initial data u_0. Then $P_1 = P$, although the mapping $h(\cdot, \lambda) = Id - P_\lambda$ is destined to fail as a possible homotopy since $h(\cdot, 0) \equiv 0$. However, a slight modification of this idea will prove useful for finding sufficient conditions for the existence of solutions. Indeed, we may define

$$h(u_0, \lambda) = \begin{cases} \frac{u_0 - P_\lambda(u_0)}{\lambda} & \text{if } \lambda \neq 0, \\ f(0, u_0) & \text{if } \lambda = 0. \end{cases}$$

A simple computation shows that h is continuous since, for $\lambda > 0$, if u is the corresponding solution of the initial value problem, then

$$h(u_0, \lambda) = \frac{u(0) - u(\lambda)}{\lambda} = -u'(\xi) = f(\xi, u(\xi))$$

for some $\xi \in (0, \lambda)$, and hence

$$h(u_0^n, \lambda_n) \to f(0, u_0)$$

for any $(u_0^n, \lambda_n) \to (u_0, 0)$. In consequence, if $h(u_0, \lambda) \neq 0$ for all $u_0 \in \partial\Omega$ and all λ, then the degree of $h(\cdot, \lambda)$ is defined and

$$deg(Id - P, \Omega, 0) = deg(h(\cdot, 1), \Omega, 0) = deg(f(0, \cdot), \Omega, 0). \qquad (5.11)$$

Summarizing, we have obtained the following result, which was proven more generally in [59] (see also [85], in case you don't read Russian).

Proposition 5.1. *Let $f : [0, 1] \times \mathbb{R}^n \to \mathbb{R}^n$ be continuous and locally Lipschitz in u, and let $\Omega \subset \mathbb{R}^n$ be open and bounded such that the solutions of the initial value problem that start in $\overline{\Omega}$ are defined up to $t = 1$. Assume, moreover, the following conditions:*

1. *$P_\lambda(u_0) \neq u_0$ for $\lambda \in (0, 1)$ and $u_0 \in \partial\Omega$.*
2. *$f(0, u_0) \neq 0$ for $u_0 \in \partial\Omega$.*
3. *$deg(f(0, \cdot), \Omega, 0) \neq 0$.*

Then the periodic problem (5.10) admits at least one solution.

5.2 To Infinity and Beyond: The Leray–Schauder Degree

In this section, we shall give an extension of the Brouwer degree to Banach spaces. As the reader might already suspect, this is not possible for arbitrary continuous functions. For example, as shown in Exercises 4.5 and 5.13, in infinite-dimensional Banach spaces there may exist continuous homotopies between the identity function of the unit ball and a constant. However, as Leray and Schauder showed in [71], the theory of topological degree can be extended for compact perturbations of the identity, that is, operators $F = I - K$ with K compact. Later on, other generalizations were developed, but they are beyond the scope of this work (for a history and account of some extensions see [84]).

The idea of the Leray–Schauder construction relies on the fact, mentioned in Remark 4.1, that if $\Omega \subset E$ is bounded and open and $K : \overline{\Omega} \to E$ is compact, then for arbitrary $\varepsilon > 0$ there exists $K_\varepsilon : \overline{\Omega} \to E$ continuous such that $I_m(K_\varepsilon)$ is contained in a finite-dimensional subspace V_ε and $\|K(x) - K_\varepsilon(x)\| < \varepsilon$ for all $x \in \overline{\Omega}$ Then, the degree of $I - K$ may be defined as the Brouwer degree of its restriction to V_ε. With this aim we shall need, first, to extend the Brouwer degree to arbitrary finite-dimensional vectors and, second, to prove that if ε is small enough, then the definition does not depend on the choice of K_ε.

Definition 5.5. Let $V \simeq \mathbb{R}^n$ be a normed space, let $\Omega \subset V$ be open and bounded, and let $\psi : V \to \mathbb{R}^n$ be an isomorphism. For $f : \overline{\Omega} \to V$ continuous with $f \neq p$ on $\partial\Omega$, we define

$$deg(f,\Omega,p) := deg(\psi \circ f \circ \psi^{-1}, \psi(\Omega), \psi(p)).$$

It is an easy exercise to verify that the previous definition is independent of ψ. For example, it suffices to assume that p is a regular value of f and use the chain rule.

Lemma 5.4. Let $\Omega \subset \mathbb{R}^n$ be open and bounded, let $m < n$, and let $f : \overline{\Omega} \to \mathbb{R}^m$ be continuous. Identify $\mathbb{R}^m$ with $\mathbb{R}^m \times \{0\} \subset \mathbb{R}^n$, and let $g : \overline{\Omega} \to \mathbb{R}^n$ be given by $g(x) = x - f(x)$. If $p \in \mathbb{R}^m$ is such that $p \notin g(\partial\Omega)$, then

$$deg(g,\Omega,p) = deg(g|_{\overline{\Omega} \cap \mathbb{R}^m}, \Omega \cap \mathbb{R}^m, p). \tag{5.12}$$

Proof. First, observe that $g(\overline{\Omega} \cap \mathbb{R}^m) \subset \mathbb{R}^m$, so the right-hand-side term of (5.12) is well defined. It suffices to prove the result for f of class C^1 and p a regular value of g. If $g(x) = p \in \mathbb{R}^m$, then $x = f(x) + p \in \mathbb{R}^m$, and hence $g^{-1}(p) \subset \Omega \cap \mathbb{R}^m$. Thus, the preimage of p by the function g coincides with its preimage by its restriction to $\overline{\Omega} \cap \mathbb{R}^m$. On the other hand, writing

$$g(x) = (x_1 - f_1(x), \ldots, x_m - f_m(x), x_{m+1}, \ldots, x_n)$$

we obtain, for $x \in \Omega \cap \mathbb{R}^m$,

$$Dg(x) = \begin{pmatrix} I_m - \left(\frac{\partial f_i}{\partial x_j}(x)\right)_{1 \leq i,j \leq m} & -\left(\frac{\partial f_i}{\partial x_j}(x)\right)_{m < j \leq n} \\ 0 & I_{n-m} \end{pmatrix},$$

where I_j denotes the identity matrix of $\mathbb{R}^{j \times j}$. Moreover, the differential matrix of $g|_{\overline{\Omega} \cap \mathbb{R}^m}$ at x is simply the upper left block of $Dg(x)$, so the result follows. $\square$

As a final preliminary step for the definition of the Leray–Schauder degree, we observe that if $F = I - K$ as before, then $dist(0, F(\partial\Omega)) > 0$. This is obviously false when F is just continuous and $dim(E) = \infty$, but the compactness plays a crucial role, as seen in the following lemma.

Lemma 5.5. Let $\Omega \subset E$ be open and bounded, and let $K : \overline{\Omega} \to E$ be compact such that $Kx \neq x$ for all $x \in \partial\Omega$. Then

$$\inf_{x \in \partial\Omega} \|x - K(x)\| > 0.$$

Proof. Suppose that $K(x_n) - x_n \to 0$ for some $x_n \in \partial\Omega$. Passing to a subsequence if necessary, we may assume that Kx_n converges to some $x \in \partial\Omega$. Hence x_n converges to the same x and $Kx = x$, a contradiction. $\square$

We are ready to define the degree for compact perturbations of the identity. By the translation property, it suffices to define it for $p = 0$.

Definition 5.6. Let $\Omega \subset E$ be open and bounded, and let $K : \overline{\Omega} \to E$ be compact with $Kx \neq x$ for $x \in \partial\Omega$. Set $\varepsilon = \inf_{x \in \partial\Omega} \|x - Kx\|$, and define

$$deg_{LS}(I - K, \Omega, 0) = deg((I - K_\varepsilon)|_{V_\varepsilon}, \Omega \cap V_\varepsilon, 0),$$

where K_ε is an ε-approximation of K with $Im(K_\varepsilon) \subset V_\varepsilon$ and $dim(V_\varepsilon) < \infty$.

To see that the Leray–Schauder degree is well defined, let K_ε and $\tilde{K}_\varepsilon$ be ε-approximations of K with ranges in the finite-dimensional subspaces V_ε and $\tilde{V}_\varepsilon$, respectively. By Lemma 5.4, if $V := V_\varepsilon + \tilde{V}_\varepsilon$, then defining the appropriate isomorphisms with the corresponding euclidean spaces yields

$$deg((I - K_\varepsilon)|_{V_\varepsilon}, \Omega \cap V_\varepsilon, 0) = deg((I - K_\varepsilon)|_V, \Omega \cap V, 0),$$

$$deg((I - \tilde{K}_\varepsilon)|_{\tilde{V}_\varepsilon}, \Omega \cap \tilde{V}_\varepsilon, 0) = deg((I - \tilde{K}_\varepsilon)|_V, \Omega \cap V, 0).$$

Moreover, the homotopy

$$h(x, \lambda) := \lambda(I - K_\varepsilon) + (1 - \lambda)(I - \tilde{K}_\varepsilon)$$

does not vanish on $\partial(\Omega \cap V) \subset \partial\Omega \cap V$; indeed, if $h(x, \lambda) = 0$ for some $x \in \partial\Omega$, then

$$x = \lambda K_\varepsilon x + (1 - \lambda)\tilde{K}_\varepsilon x,$$

and hence

$$\|x - Kx\| \leq \lambda \|K_\varepsilon x - Kx\| + (1 - \lambda)\|\tilde{K}_\varepsilon x - Kx\| < \varepsilon,$$

a contradiction. We conclude that

$$deg((I - K_\varepsilon)|_V, \Omega \cap V, 0) = deg((I - \tilde{K}_\varepsilon)|_V, \Omega \cap V, 0).$$

Summarizing, we have constructed a function deg that assigns to each operator $F = I - K : \overline{\Omega} \to E$ with K compact and $0 \notin F(\partial\Omega)$ an integer with the following properties (proofs are left as an exercise):

1. *Normalization:* $deg(Id, \Omega, 0) = \begin{cases} 1 & \text{if } 0 \in \Omega, \\ 0 & \text{if } 0 \notin \Omega. \end{cases}$
2. *Solution:* If $deg(F, \Omega, 0) \neq 0$, then F vanishes in Ω.
3. *Excision:* If $\Omega_1 \subset \Omega$ is open and F does not vanish in $\overline{\Omega} \setminus \Omega_1$, then $deg(F, \Omega, p) = deg(F, \Omega_1, p)$.
4. *Homotopy invariance:* If $K : \overline{\Omega} \times [0, 1] \to E$ is compact such that $K(x, \lambda) \neq x$ for $x \in \partial\Omega$ and $\lambda \in [0, 1]$, then $deg(I - K(\cdot, \lambda), \Omega, 0)$ is constant.

Remark 5.3. * In many textbooks, the latter property is stated for the more general situation that K is a *homotopy of compact operators*, that is:

1. $K(\cdot, \lambda)$ is compact for every $\lambda \in [0,1]$.
2. For each $\varepsilon > 0$ there exists $\delta > 0$ such that $\|K(x, \lambda_1) - K(x, \lambda_2)\| < \varepsilon$ whenever $x \in \overline{\Omega}$ and $|\lambda_1 - \lambda_2| < \delta$.

In this case, it is easy to prove that the assumption $K(x, \lambda) \neq x$ on $\partial\Omega$ implies

$$\inf_{x \in \partial\Omega, \lambda \in [0,1]} \|x - K(x, \lambda)\| := r > 0,$$

and hence, for fixed λ_0, there exists an $\frac{r}{2}$-approximation K of $K(\cdot, \lambda_0)$. Thus, K is an r-approximation of $K(\cdot, \lambda)$ for λ close to λ_0. By definition of the Leray–Schauder degree, $deg(I - K(\cdot, \lambda), \Omega, 0) = deg(I - K, \Omega, 0)$, and we deduce that the degree is locally constant (and hence constant) with respect to λ.

Remark 5.4. In accordance with the final paragraph of Sect. 5.1, homotopy invariance admits also continuous changes in the domain. That is, if $U \subset E \times [0,1]$ is open and bounded and $U_\lambda := \{x \in E : (x, \lambda) \in U\}$, then for any $K : \overline{U} \to E$ compact such that $x \neq K(x, \lambda)$ for $(x, \lambda) \in \partial U$ the degree $deg(I - K(\cdot, \lambda), U_\lambda, 0)$ is independent of λ.

Much more can be said about the degree; see, for example, [27, 30, 35] or [72]. Here, we have restricted ourselves to those properties that shall be used in our applications.

And talking about applications, the first of them is, undoubtedly, Schauder's theorem. The proof can be done exactly in the same way as we did for Brouwer when, for example, C is a closed ball centered at 0 of the Banach space E: if $Tx = x$ for some $x \in \partial C$, then there is nothing to prove; otherwise, the homotopy $h(x, \lambda) = x - \lambda Tx$ does not vanish on ∂C, and hence its degree is constantly equal to 1. However, if C is an arbitrary convex closed subset of E, it might happen that its interior is empty, and hence degree theory does not apply. In such an unfortunate case, there is a theorem that comes to our aid: we refer to *Dugundji's theorem* (see Theorem 5.10 in the appendix), which allows us to extend T to a compact operator $\tilde{T} : \overline{B_R(0)} \to C$ for some ball $B_R(0)$ containing C. Then $\tilde{T}$ has a fixed point in C, which is also a fixed point of T.

There is also a beautiful result due to Krasnoselskii (e.g., [27, 109]), which establishes the existence of a fixed point for a compact operator in a cone. It can be proven using Schauder's theorem, but degree theory provides a slightly shorter proof.

Theorem 5.7. *(Krasnoselskii) Let E be a Banach space, and let $C \subset E$ be a closed cone. Let $T : C \to C$ be compact and assume, for some positive $r \neq s$, that*

1. *$Tx - x \notin C$ for $\|x\| = r$,*
2. *$x - Tx \notin C$ for $\|x\| = s$.*

Then T has at least one fixed point.

The geometric interpretation of the Krasnoselskii theorem with $r > s$ is shown in the following figure:

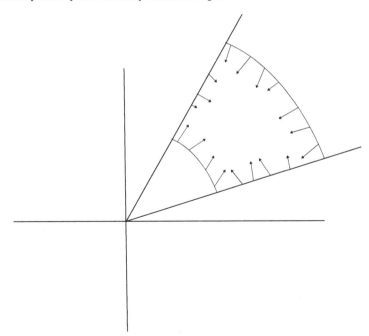

The cases $r < s$ and $r > s$ are usually referred to respectively as the *expansive form* and the *compressive form*: in the first case, the operator is outwardly pointing over the boundary of the annular sector and inwardly pointing in the second case. There are many different versions and extensions of this result; for example, in [58] the conditions are replaced by

1' $\|Tx\| \le \|x\|$ for $\|x\| = r$,
2' $\|x\| \le \|Tx\|$ for $\|x\| = s$.

Proof of Theorem 5.7. Let us assume that $r < s$; the other case is similar and left as an exercise. By Dugundji's theorem, T can be extended to a compact operator, still denoted T, such that $Im(T) \subset co(T(C)) \subset C$, where "co" stands for the convex hull. Note that if x is a fixed point of T, then $x \in C$, and from the hypotheses, $\|x\| \ne r, s$. Thus, the degree of $I - T$ is well defined on $B_r(0)$ and $B_s(0)$ and, moreover,

$$deg(I - T, B_s(0), 0) = deg(I - T, B_r(0), 0) + deg(I - T, B_s(0) \backslash \overline{B_r(0)}, 0).$$

We claim that $deg(I - T, B_r(0), 0) = 1$ and $deg(I - T, B_s(0), 0) = 0$; consequently, the latter degree is -1 and, in particular, $I - T$ vanishes in $B_s(0) \backslash \overline{B_r(0)}$. Furthermore, if x is a zero of $I - T$, then x is a fixed point of T; thus, $x \in C$, and the proof is complete.

To prove the claim, define firstly the homotopy $h(x, \lambda) = x - \lambda Tx$ for $\lambda \in [0, 1]$. If $h(x, \lambda) = 0$, then $\lambda Tx - x = 0 \in C$, and hence $x \le \lambda Tx \le Tx$, that is, $Tx - x \in C$. This implies that $\|x\| \ne r$; in other words, $I - T$ is homotopic to I on $\overline{B_r(0)}$, and thus $deg(I - T, B_r(0), 0) = 1$.

Finally, observe that $T(\overline{B_s(0)})$ is a bounded set, and hence $(I - T)(\overline{B_s(0)})$ is also bounded; thus, we may take $y \in C$ such that $x - Tx \neq y$ for all $x \in \overline{B_s(0)}$. Next, define the homotopy $h(x, \lambda) = x - K_\lambda x$, where $K_\lambda x := Tx + \lambda y$, which is clearly a compact operator. If $h(x, \lambda) = 0$, then $x - Tx = \lambda y \in C$, so we deduce that $\|x\| \neq s$. Hence

$$deg(I - T, B_s(0), 0) = deg(I - T - y, B_s(0), 0) = 0$$

since $I - T - y$ does not vanish in $B_s(0)$.

If the first condition is replaced by $1'$, then it may be assumed that T has no fixed points with norm equal to r and repeat the preceding proof. If the second condition is replaced by $2'$ and T has no fixed points with norm equal to s, then, first, observe that $0 \notin co(T(\partial B_s(0)))$; indeed, $T(\partial B_s(0)) \in C\backslash\{0\}$, and if $0 = \sum_{j=1}^N a_j x_j$ for some $a_j > 0$ and $x_j \in C\backslash\{0\}$, then $-a_N x_N = \sum_{j=1}^{N-1} a_j x_j \in C$, a contradiction. Moreover, the degree depends only on the values the function takes at the boundary, so using Dugundji's theorem it may be assumed that $\|Tx\| \geq c > 0$ for all $x \in \overline{B_s(0)}$. Let $h(x, \lambda) = x - (1 + \lambda)Tx$, then $h(x, \lambda) \neq 0$ for $\|x\| = s$ and $\lambda \geq 1$ and, moreover, h does not vanish in $\overline{B_s(0)}$ for $\lambda > \frac{s}{c} - 1$, which implies $deg(h(\cdot, 0), B_s(0), 0) = 0$.

We end this chapter with an immediate application of the Leray–Schauder degree theory to an abstract nonresonant problem $Lu = Nu$, as we did in Example 4.3. Assume that L has a compact inverse and N maps bounded sets into bounded sets; then the problem may be written as $u = Ku$, where $K = L^{-1}N : E \to E$ is compact. In this case, a standard homotopy would be $h(u, \lambda) = u - \lambda Ku$; thus, the properties of the Leray–Schauder degree imply that the problem has a solution, provided that we are able to find an open bounded $\Omega \subset E$ such that $0 \in \Omega$ and $u \neq \lambda Ku$ for $u \in \partial\Omega$ and $\lambda \in (0, 1)$. Indeed, simply observe that we may assume that $u \neq Ku$ for $u \in \partial\Omega$ (since, otherwise, K already has a fixed point). Moreover, $h(u, 0) \neq 0$ for $u \in \overline{\Omega}$ since $0 \in \Omega$, and hence

$$deg(I - K, \Omega, 0) = deg(I, \Omega, 0) = 1.$$

The previous abstract result is readily adapted to our (nonresonant) model equation

$$u''(t) = f(t, u(t)), \qquad u(0) = u(1) = 0, \tag{5.13}$$

with $f : [0, 1] \times \mathbb{R}^n \to \mathbb{R}^n$ continuous. If for all $\lambda \in (0, 1)$ the problem

$$u''(t) = \lambda f(t, u(t)), \qquad u(0) = u(1) = 0 \tag{5.14}$$

has no solutions on $\partial\Omega$ for some open bounded $\Omega \subset C([0, 1])$ containing 0, then (5.13) has a solution in $\overline{\Omega}$. This explains many of the results obtained in the previous chapters: as soon as one gets a priori bounds for (5.14), the problem is solved. Indeed, if there is a constant R such that $\|u\|_\infty < R$ for any u solution of (5.14) with $0 < \lambda < 1$, then it suffices to take $\Omega = B_R(0)$. This is the case, for example, when f is sublinear, or nondecreasing in the second coordinate. The same is true if we assume the Hartman condition

$$f(t,u) \cdot u \geq 0 \qquad \text{for } |u| = R. \tag{5.15}$$

Here, it is not guaranteed that the homotopy will not vanish at any point of the boundary, but one may solve the problem assuming firstly that the inequality in (5.15) is strict and then using an approximation argument. Sometimes a slightly more sophisticated method is required: for example, assume as before that f is continuous and satisfies (5.15), but consider now the first-order problem

$$u'(t) = f(t,u(t)), \qquad u(0) = u(1).$$

Here the operator $Lu := u'$ is noninvertible, so the system cannot be directly stated as a fixed point problem. However, the simple trick of considering the equivalent equation

$$u'(t) - u(t) = f(t,u(t)) - u(t)$$

allows us to define $K : C([0,T]) \to C([0,1])$ by $Kv = u$, where u is the unique solution of the problem

$$u'(t) - u(t) = f(t,v(t)) - v(t), \qquad u(0) = u(1).$$

If $u = \lambda Ku$ for some $\lambda \in (0,1)$, then

$$u'(t) = \lambda f(t,u(t)) + (1 - \lambda)u(t).$$

Suppose that $\|u\|_\infty = R$, and fix $t_0 \in [0,1)$ such that the absolute maximum of the function $\phi(t) := \frac{|u(t)|^2}{2}$ is achieved at t_0; then

$$0 \geq \phi'(t_0) = u'(t_0) \cdot u(t_0) = \lambda f(t_0, u(t_0)) \cdot u(t_0) + (1 - \lambda)R^2 > 0,$$

a contradiction. Thus, the problem has a solution in $\overline{B_R(0)} \subset C([0,1])$. The reader may verify that a similar result holds if (5.15) is reversed, that is, $f(t,u) \cdot u \leq 0$ for $|u| = R$. Also, the Hartman condition is sufficient for proving the existence of periodic solutions of the second-order problem $u''(t) = f(t,u(t))$. Again, the operator u'' is noninvertible, but we may imitate the previous example and consider $Kv := u$ defined as the unique solution of

$$u''(t) - u(t) = f(t,v(t)) - v(t), \qquad u(0) - u(1) = u'(0) - u'(1) = 0.$$

If $u = \lambda Ku$ for some $\lambda \in (0,1)$, $\|u\|_\infty = R$ and ϕ and t_0 are defined as before, then it is easily seen that $0 \geq \phi''(t_0) \geq u''(t_0) \cdot u(t_0) > 0$, a contradiction.

The last examples constitute the most elementary cases of the so-called *continuation methods*. More general continuation theorems and specific applications to boundary value problems and resonant problems shall be given in the next chapter.

Appendix

Many important results in analysis have been used in our construction of the Brouwer degree. Most of them are very well known and require no presentation here, for example, the divergence theorem, the change of variables theorem, or the Stone–Weierstrass theorem. But others, still very important, are less famous and deserve a few paragraphs. The first one is usually referred as a lemma, although it is very strong and beautiful. We shall only prove the case that has been used in this work (as mentioned, a proof for the general case can be found, for example, in [87]).

Theorem 5.8. *(Sard's lemma, particular case) Let $U \subset \mathbb{R}^n$ be open and $f : U \to \mathbb{R}^n$ be a C^2 function. Then the set $\mathscr{C} \subset \mathbb{R}^n$ of critical values of f has zero Lebesgue measure.*

Proof. By the σ-subadditivity of the Lebesgue measure, we may assume that U is a cube of side L and that f and its derivatives are bounded by some constant M. Divide U into N^n cubes of side $\frac{L}{N}$, let x be a critical point of f, and denote by C_x the small cube of side $\frac{L}{N}$ that contains x. For $y \in Cx$, we may write

$$f(y) - f(x) = Df(x)(y - x) + R(y),$$

where $|R(y)| \leq \frac{M}{2}|y - x|^2 \leq \frac{Mn}{2}\left(\frac{L}{2N}\right)^2$. Recall that x is a critical point, so the linear transform $Df(x)$ maps the cube C_x into a ball of smaller dimension and radius $M\sqrt{n}\frac{L}{2N}$; hence, the set $\{f(y) : y \in C_x\}$ is contained in a cylinder centered at $f(x)$ whose basis is $c\left(\frac{L}{N}\right)^{n-1}$ for some c and whose height is $Mn\left(\frac{L}{2N}\right)^2$. Next, observe that every critical value is an image of some critical point (and there are N^n cubes), so it follows that

$$|\mathscr{C}| \leq N^n c \left(\frac{L}{N}\right)^{n-1} M\sqrt{n}\left(\frac{L}{2N}\right)^2 \leq \frac{C}{N}$$

for some constant C. Letting $N \to \infty$, we conclude that $|\mathscr{C}| = 0$. $\square$

In addition, several extension theorems have been used in different applications. The first of them is very well known, and the proof may be found in all basic textbooks on metric spaces.

Theorem 5.9. *(Tietze) Let X be a metric space, let $A \subset X$ be closed, and let $f : A \to \mathbb{R}$ be continuous. Then there exists a continuous function $F : X \to \mathbb{R}$ such that $F|_A = f$.*

As an immediate consequence, the result is still valid for $f : A \to \mathbb{R}^n$; furthermore, the extension F can be chosen in such a way that $F(A) \subset co(f(A))$, where, as mentioned, "co" stands for the convex hull. But in many cases a more general result is needed for a function with values in an arbitrary Banach space. This generalization is due to Dugundji. Its proof was first published in [33] and can be found in [27], among other texts.

Theorem 5.10. *(Dugundji) Let X be a metric space, let $A \subset X$ be closed, and let $f : A \to Y$ be continuous, where Y is a Banach space. Then there exists a continuous function $F : X \to co(f(A))$ such that $F|_A = f$.*

Remark 5.5. In particular, if $\overline{f(A)}$ is compact, then its convex hull is compact, and hence F is a compact operator.

The preceding theorems have several useful variants. In particular, when f does not vanish, it might happen that $0 \in co(f(A))$, but still, if the domain has a smaller dimension than the codomain, then there is enough "room" to construct a nonvanishing extension. More precisely:

Lemma 5.6. *Let $K \subset \mathbb{R}^n$ be compact and $f : K \to \mathbb{R}^m\setminus\{0\}$ continuous, with $m > n$. Then there exists a continuous extension $F : \mathbb{R}^n \to \mathbb{R}^m\setminus\{0\}$.*

Proof. First, observe that if $K = \overline{B_R(0)}$, then f can be extended by

$$F(x) = \begin{cases} f(x) & |x| \leq R, \\ f\left(R\frac{x}{|x|}\right) & |x| > R. \end{cases}$$

Thus, for the general case it suffices to show that f can be extended to some closed ball $\overline{B}$ centered at 0.

Let $r = \min_{x \in K}|f(x)|$, and take $g \in C^1(\overline{B}, \mathbb{R}^m)$ such that $|g(x) - f(x)| < \frac{r}{4}$ for $x \in K$. By Sard's lemma, there exists $y \notin Im(g)$ such that $|y| < \frac{r}{4}$. Let

$$\sigma(t) = \begin{cases} \frac{2t}{r} & t \leq \frac{r}{2}, \\ 1 & t > \frac{r}{2}, \end{cases}$$

and define $\phi(x) = \frac{g(x)-y}{\sigma(g(x)-y)}$.

It follows by simple computation that $|\phi(x)| \geq \frac{r}{2}$ for all $x \in \overline{B}$; furthermore, if $x \in K$, then

$$|g(x) - y| \geq |f(x)| - |g(x) - f(x)| - |y| > \frac{r}{2}.$$

Hence $\phi(x) = g(x) - y$ and, in particular, $|\phi(x) - f(x)| \leq |f(x) - g(x)| + |y| < \frac{r}{2}$. Thus, $co(Im(\phi - f)) \subset B_{r/2}(0)$, so by Dugundji's theorem there exists $T : \overline{B} \to \mathbb{R}^m$ continuous extending $\phi - f$ such that $|T(x)| \leq \max_{z \in K}|\phi(z) - f(z)| < \frac{r}{2}$ for all $x \in \overline{B}$. Then $F : \overline{B} \to \mathbb{R}^m$ given by $F(x) := \phi(x) - T(x)$ is continuous and satisfies $|F(x)| \geq |\phi(x)| - |T(x)| \geq \frac{r}{2} - |T(x)| > 0$ for all $x \in \overline{B}$. Clearly, F is an extension of f. $\square$

As corollaries, we deduce the next two lemmas, which were used in our first proof of Borsuk's theorem.

Lemma 5.7. *Let $V \subset \mathbb{R}^n$ be open bounded and symmetric such that $0 \notin \overline{V}$, let $m > n$, and let $f : \partial V \to \mathbb{R}^m\setminus\{0\}$ be continuous and odd. Then there exists a continuous odd extension $F : \overline{V} \to \mathbb{R}^m\setminus\{0\}$.*

Proof. We shall proceed by induction. The case $n = 1$ is left as an exercise. Next, assume the lemma is true for $n - 1$, and let $V_0 = V \cap \mathbb{R}^{n-1}$. Then $\partial V_0 \subset \partial V \cap \mathbb{R}^{n-1}$ and $f|_{\partial V_0}$ can be extended to a nonvanishing odd continuous function (still denoted f) on $\overline{V_0}$. Now define $S^+ := \{x \in \mathbb{R}^n : x_n \geq 0\}$, $K = \overline{V_0} \cup (\partial V \cap S^+)$. Using the preceding lemma, f is extended to a nonvanishing continuous function on $\overline{V} \cap S^+$, which in turn can be extended to $\overline{V}$ by $f(x) := -f(-x)$ for $x \notin S^+$. $\square$

Lemma 5.8. *Let $V \subset \mathbb{R}^n$ be open bounded and symmetric such that $0 \notin \overline{V}$, and let $f : \partial V \to \mathbb{R}^n \backslash \{0\}$ be continuous and odd. Given a hyperplane H passing through the origin, there exists a continuous odd extension $F : \overline{V} \to \mathbb{R}^n$ such that $F(x) \neq 0$ for $x \in \overline{V} \cap H$.*

Proof. Let $V_0 = V \cap H$; then $\partial V_0 \subset \partial V \cap \mathbb{R}^n$, and from the previous lemma, f can be extended to a nonvanishing continuous odd function on $\overline{V_0}$. By Tietze's theorem, this function can be extended continuously to $\overline{V} \cap S^+$ and thereafter to $\overline{V}$, as before. $\square$

Problems

Brouwer Degree

5.1. Let $f : \overline{B_1(0)} \to \mathbb{R}^n$ be continuous such that $f(\partial B_1(0)) \subsetneq \partial B_1(0)$. Prove that $deg(f, B_1(0), 0) = 0$.

5.2. Let $\Omega \subset \mathbb{R}^n$ be open and bounded, $f : \overline{\Omega} \to \mathbb{R}^n$ continuous with $0 \notin f(\partial\Omega)$, and $g : U \to V \subset \mathbb{R}^n$ a diffeomorphism, where U is an open connected neighborhood of $f(\overline{\Omega})$. Prove that if $0 \notin g \circ f(\partial\Omega)$, then

$$deg(g \circ f, \Omega, 0) = sgn(Jg(y))deg(f, \Omega, 0)$$

for any $y \in U$. In particular, $deg(-f, \Omega, 0) = (-1)^n deg(f, \Omega, 0)$.

Note: the previous equality is an elementary case of the well-known *product formula* (e.g., [30, 35, 72]).

5.3. Let $B = \{z \in \mathbb{C} : |z| < 1\}$, and let $f : \overline{B} \to \mathbb{C}$ be a nonconstant analytic function such that $f(\partial B) \subset \partial B$. Prove that

$$\sum_{n \geq 1} n \left(\frac{|f^{(n)}(0)|}{n!} \right)^2 \in \mathbb{N}.$$

Remark 5.6. According to the well-known minimum modulus principle, it follows that f has at least one zero in B. Can you see any relation between the number of zeros and the previous sum?

5.4. Prove the following theorem (Rothe): Let $\Omega \subset \mathbb{R}^n$ be open and bounded, and let $f : \overline{\Omega} \to \mathbb{R}^n$ be continuous. Suppose there exists $x_0 \in \Omega$ such that $f(x) - x_0 \neq t(x - x_0)$ for all $t > 1$ and all $x \in \partial\Omega$. Then f has a fixed point.

5.5. Prove the following generalization of the fundamental theorem of algebra. Let $P(z, \bar{z})$ be a polynomial of degree less than or equal to n. Then the equation $z^n = P(z, \bar{z})$ has at least one solution. Is it true that it has at most n solutions?

5.6. 1. Let $V \subset \mathbb{R}^n$ be open, bounded, and symmetric, and let $\varphi : \overline{V} \to \mathbb{R}$ be continuous and odd. Prove that for any $\varepsilon > 0$ there exists $\psi : \overline{V} \to \mathbb{R}$ of class C^1 and odd such that $\|\varphi - \psi\|_\infty < \varepsilon$.
   Hint: extend φ to a continuous odd function in $\mathbb{R}^n$ with compact support and take an even smooth function $\phi : \mathbb{R}^n \to \mathbb{R}$ with compact support and $\int_{\mathbb{R}^n} \phi(x)\,dx = 1$. Prove that the convolution $\varphi * \phi_\varepsilon$, where $\phi_\varepsilon(x) = \varepsilon^{-n}\phi(\frac{x}{\varepsilon})$, is odd and converges uniformly to φ. An alternative (and much simpler!) proof follows by taking an arbitrary C^1 function $\tilde{\psi}$ such that $\|\varphi - \tilde{\psi}\|_\infty < \varepsilon$ and then defining $\psi(x) := \frac{\tilde{\psi}(x) - \tilde{\psi}(-x)}{2}$.
2. Let V be as before, and let $f : \overline{V} \to \mathbb{R}^n$ be continuous and odd. Prove that for any $\varepsilon > 0$ there exists $g : \overline{V} \to \mathbb{R}$ of class C^1 and odd such that $\|f - g\|_\infty < \varepsilon$ and $Dg(0)$ is invertible.
   Hint: using 1, approximate f by an odd C^1 function $\tilde{f}$. Next, write $\tilde{f}(x) = D\tilde{f}(0)x + R(x)$ and use the fact that the set of invertible matrices is dense in $\mathbb{R}^{n\times n}$.

5.7. Let $f : S^n \to S^n$ be continuous with $n \in \mathbb{N}$ even. Prove that, for some x, either $f(x) = x$ or $f(x) = -x$.

5.8. Let $G : [0,1] \times \mathbb{R}^n \to \mathbb{R}^n$ be continuous, sublinear, and locally Lipschitz in x, and let P be the Poincaré operator associated to the problem $x'(t) + x(t) = G(t, x(t))$. Prove that P is defined in $\mathbb{R}^n$ and $deg(I \pm P, B_R(0), 0) = 1$ for any sufficiently large R.

5.9. Let $F : \mathbb{R}^n \to \mathbb{R}$ be an even C^2 function. Prove that the antiperiodic problem

$$\begin{cases} x'(t) = \nabla F(x(t)) + p(t) \\ \quad x(0) + x(1) = 0 \end{cases}$$

has at least one solution for any continuous $p : [0,1] \to \mathbb{R}^n$. Compare with Exercise 4.7.
   Hint: because $F(x) - \frac{|x|^2}{2}$ is also an even C^2 function, it suffices to prove that the equation $x'(t) + x(t) = \nabla F(x(t)) + p(t)$ has an antiperiodic solution. Multiply by $x'(t)$ and integrate over $[0,1]$ to prove that any solution x satisfies $\|x'\|_{L^2} \leq \|p\|_{L^2}$. On the other hand, write $x(t) = x(0) + \int_0^t x'(s)\,ds = x(1) - \int_t^1 x'(s)\,ds$, and deduce that $2|x(t)| \leq \|x'\|_{L^2} \leq \|p\|_{L^2}$ for all t. Thus, it may be assumed that ∇F is bounded, and the result follows from the previous exercise.

5.10. Let $F : \mathbb{R}^n \to \mathbb{R}$ be a C^1 mapping such that

$$\lim_{|x| \to \infty} F(x) \to +\infty \quad \text{and} \quad \nabla F(x) \neq 0 \text{ for } |x| \geq R.$$

Prove that $deg(\nabla F, B_r(0), 0) = 1$ for all r sufficiently large.

    Hint: consider the autonomous equation $u'(t) + \nabla F(u(t)) = 0$ with initial value $u(0) = u_0$, and prove that if u is a solution, then $F(u(t)) = F(u_0) - \int_0^t |\nabla F(u(s))|^2 \, ds$. Moreover, deduce that u is defined for all $t > 0$ and, if $|u_0| \geq R$, then $F(u(t)) < F(u_0)$ for all $t > 0$ and verify as in (5.11) that $deg(\nabla F, B_r(0), 0) = deg(Id - P_\lambda, B_r(0), 0)$ for all r large enough and all $\lambda > 0$. Finally, prove the existence of $\lambda, r > 0$ such that if $|u_0| = r$, then $|P_\lambda(u_0)| < r$ and complete the proof.

5.11. Let $f : [0,1] \times \mathbb{R}^n \to \mathbb{R}^n$ be continuous and locally Lipschitz in u. Assume there exists a bounded open $\Omega \subset \mathbb{R}^n$ containing 0 such that

1. For all $\lambda \in [0,1]$ and all $v_0 \in \overline{\Omega}$, the solution u_{λ,v_0} of the initial value problem

$$\begin{cases} u''(t) = \lambda f(t, u(t), u'(t)) \ t \in (0,1) \\ u(0) = 0, \quad u'(0) = v_0 \end{cases}$$

   is defined over $[0,1]$;
2. If $v_0 \in \partial\Omega$, then $u_{\lambda,v_0}(1) \neq 0$.

    Prove that the Dirichlet problem

$$\begin{cases} u''(t) = \lambda f(t, u(t), u'(t)) \ t \in (0,1) \\ u(0) = u(1) = 0 \end{cases}$$

has at least one solution u with $u'(0) \in \Omega$.

Leray–Schauder Degree

5.12. * Let $\hat{x}, r \in l^2 := \{(x_n)_{n \in \mathbb{N}} \subset \mathbb{R}^{\mathbb{N}} : \sum_{n=1}^\infty x_n^2 < \infty\}$, and let $\mathscr{C}$ be the so-called *Hilbert cube* defined by

$$\mathscr{C} := \{x \in l^2 : |x_i - \hat{x}_i| \leq |r_i| \quad \text{for all } i\}.$$

Prove that if $F : \mathscr{C} \to l^2$ is continuous and

$$F_i(y) \leq 0 \leq F_i(z)$$

for every $y, z \in \mathscr{C}$ such that $y_i = \hat{x}_i - r_i$, $z_i = \hat{x}_i + r_i$, then F has at least one zero.

5.13. 1. Give an example of a continuous retraction $r : \overline{B} \to \partial B$, where B is a ball of a Banach space E.

2. Using the previous example, find $h : \overline{B} \times [0,1] \to E\setminus\{0\}$ continuous such that $h(x,0) = c$ and $h(x,1) = x$ for all $x \in \partial B$. Does this contradict the homotopy invariance of the Leray–Schauder degree?

Remark 5.7. Another example of such a homotopy was given in Exercise 4.5. According to the usual terminology, this asserts that the sphere of E is *contractible*.

5.14. Let E be a Banach space and let $K : E \to E$ be compact such that $\frac{Kx}{\|x\|} \to 0$ as $\|x\| \to \infty$. Prove that the equation $x = Kx + y$ has at least one solution for arbitrary $y \in E$.

5.15. Extend the Borsuk theorem and its consequences for an odd mapping $F = I - K$, where K is a compact operator defined in a symmetric subset of a Banach space. In particular, deduce an infinite-dimensional version of Theorem 5.6.

5.16. Prove the *Fredholm alternative*: if X is a Banach space and $K : X \to X$ is linear and compact, then for any $\mu \neq 0$ either μ is an eigenvalue of K or $\mu - K$ is an isomorphism. More generally, if K is compact and is homogeneous of degree 1, that is, $K(rx) = rKx$ for all $r \in \mathbb{R}$ and $Kx \neq \mu x$ for all x, then $\mu - K$ is surjective.
   Hint: it may be assumed without loss of generality that $\mu = 1$. If $F = I - K$ is injective, then prove that $\deg(I - K, B_R(0), 0) \neq 0$ for all $R > 0$.

5.17. Let $K, L : E \to E$ be compact operators, with L linear. Suppose that

$$\|Kx - Lx\| < \|x - Lx\|$$

for all x such that $\|x\| = R$. Prove that $\deg(I - K, B_R(0), 0)$ is odd.

5.18. Let $g : [0,1] \times \mathbb{R}^n \to \mathbb{R}^n$ be continuous with $g(\cdot, 0) \equiv 0$, and let

$$X := \{u \in C[0,1] : u(0) = u(1) = 0, \|u\|_\infty < r\}.$$

Prove that if $u'' + g(\cdot, u) \neq v'' + g(\cdot, v)$ for all $u, v \in X \cap C^2([0,1])$ such that $u \neq v$, then there exists $M > 0$ such that the problem

$$u''(t) + g(t, u(t)) = p(t), \qquad u(0) = u(1) = 0$$

has a unique solution $u \in X$ for any $p : [0,1] \to \mathbb{R}^n$ continuous with $\|p\|_\infty < M$. Furthermore, the mapping $p \mapsto u$ is continuous.

Remark 5.8. In particular, taking $g(t, u) = a(t)u$, we retrieve a classical "injectivity implies surjectivity" result: if the problem $u''(t) + a(t)u(t) = 0$ admits only the trivial solution in X, then the nonhomogeneous problem $u''(t) + a(t)u(t) = p(t)$ has a (unique) solution $u \in X$ for all p, and the mapping $p \mapsto u$ is continuous. Observe that, due to linearity, it suffices to prove the existence of solutions for p in a neighborhood of 0. One may argue that, here, the involved mappings are C^1, and hence the standard implicit function theorem can be used; however, the procedure would require verifying that the operator $u'' + au$ is an isomorphism, and this is exactly what we want to prove. More generally, the reader might try proving the following variant of the Fredholm alternative: let X, Y be Banach spaces, and let $L : D \subset X \to Y$ be linear with compact inverse $K : Y \to X$. For any $T : X \to Y$ linear and continuous, if $L + T$ is injective, then it is surjective and $(L + T)^{-1}$ is continuous.

5.19. 1. Extend Exercise 5.4 to the case of a compact operator $K : \overline{\Omega} \to E$, where $\Omega \subset E$ is open and bounded.
2. Deduce the existence of a fixed point for a compact operator $K : \overline{B_1(0)} \to E$ in the following cases:

- $K(\partial B_1(0)) \subset \overline{B_1(0)}$;
- $\|Kx - x\|^2 \geq \|Kx\|^2 - 1$ for $\|x\| = 1$;
- E is a Hilbert space and $\langle Kx, x \rangle \leq 1$ for $\|x\| = 1$.

5.20. Using the Leray–Schauder degree theory, prove the existence of solutions of the following problems:

1.
$$x'(t) = f(t, x(t)), \qquad x(t_0) = x_0$$

for $f : \mathbb{R}^{n+1} \to \mathbb{R}^n$ continuous.

2.
$$u''(t) = f(t, u(t)), \qquad u(0) = u(1) = 0$$

for $f : [0, 1] \times \mathbb{R}^n \to \mathbb{R}^n$ continuous and bounded.

5.21. * Let $g : \mathbb{R} \to \mathbb{R}$ be continuous. Prove the following assertions:

1. If there exists $R_0 > 0$ such that

$$g(-u) > 0 > g(u) \quad \text{for all } u > R_0, \tag{5.16}$$

then the periodic problem

$$u''(t) + g(u(t)) = p(t), \qquad u(0) = u(1), \quad u'(0) = u'(1)$$

has at least one solution for arbitrary $p \in C([0, 1])$ with $\overline{p} = 0$. Is the result true if inequalities (5.16) are reversed?
2. If (5.16) or the reversed inequalities hold and $c \in C([0, 1])$ is such that $c(t) \neq 0$ for all t, then the problem

$$u''(t) + c(t)u'(t) + g(u(t)) = p(t), \qquad u(0) = u(1), \quad u'(0) = u'(1)$$

has at least one solution for arbitrary $p \in C([0, 1])$ with $\overline{p} = 0$.

Chapter 6
Applications

6.1 Continuation Theorems: From Concrete to Abstract

As seen in the last chapter, the topological degree may be a powerful tool for solving problems of the type $Lu = Nu$, where L is a linear operator and N is continuous. Thus far, we have only analyzed the nonresonant case, in which L is invertible such that $K := L^{-1}N$ is compact, and used just the linear homotopy $h(u, \lambda) = u - \lambda Ku$. This is already useful and ensures the existence of solutions, provided there is a bounded neighborhood of 0 such that h does not vanish for $u \in \partial\Omega$. More precisely:

Theorem 6.1. *Let E and F be Banach spaces, and let $L : D \subset E \to F$ be linear, $N : E \to F$ continuous. Assume that L is one-to-one and $K := L^{-1}N$ is compact. Furthermore, assume there exists a bounded and open subset $\Omega \subset E$ with $0 \in \Omega$ such that the equation $Lu = \lambda Nu$ has no solutions in $\partial\Omega \cap D$ for any $\lambda \in (0, 1)$. Then the problem $Lu = Nu$ has at least one solution in $\overline{\Omega}$.*

However, a wide range of different results can be obtained allowing more general situations. Although an abstract general setting is possible, we shall start by introducing some particular examples.

6.1.1 First-Order Periodic Systems

As a first example, consider the periodic problem

$$u'(t) = f(t, u(t)), \qquad u(0) = u(1), \tag{6.1}$$

where $f : [0, 1] \times \mathbb{R}^n \to \mathbb{R}^n$ is continuous. Here, we may consider $E = F = C([0, 1])$ and $D = \{u \in C^1([0, 1]) : u(0) = u(1)\}$. Note that L is not invertible since its kernel is the set of constant functions, which may be identified with $\mathbb{R}^n$. Let us begin by an intuitive approach: first, observe that $u \in E$ solves the differential equation if and only if

$$u(t) = u(0) + Tu(t),$$

where $T : E \to E$ is the (compact) operator defined by

$$Tu(t) := \int_0^t f(s, u(s)) \, ds.$$

However, this is not an appropriate fixed point operator since the boundary condition has not yet been considered. So our challenge consists in writing such a condition also in the form of a functional equation $Pu = 0$ and, furthermore, obtain a unique equation $u = Ku$ from the previous two. With this idea in mind, just observe that, when u is differentiable, the condition $u(0) = u(1)$ may be equivalently written as $\int_0^1 u'(t) \, dt = 0$; thus, it is clear that our original problem is equivalent to the system of equations

$$u = u(0) + Tu, \qquad Pu = 0,$$

where

$$Pu := \int_0^1 f(t, u(t)) \, dt.$$

Next, observe that $Tu(0)$ is always equal to 0, which implies that both equations are satisfied at once if and only if

$$u = u(0) + Pu + Tu. \tag{6.2}$$

Indeed, if u is a solution of the original problem, then $Pu = 0$ and $u = u(0) + Tu$, so (6.2) is satisfied. Conversely, if $u = u(0) + Tu + Pu$, then evaluating both sides at $t = 0$ yields $Pu = 0$, and hence $u = u(0) + Tu$. All of this was really trivial but reduces the problem to the following one: find a zero of $Fu := u - Ku$ with $Ku :=$ $u(0) + Tu + Pu$ compact. Furthermore, the procedure gives a good clue as to how to define an appropriate homotopy: just add a λ to the last term of (6.2). We obtain the one-parameter family of problems $F_\lambda u = 0$, where

$$F_\lambda u := u - (u(0) + Pu + \lambda Tu).$$

When $\lambda > 0$, it is readily seen, as before, that $F_\lambda u = 0$ if and only if

$$u'(t) = \lambda f(t, u(t)), \qquad u(0) = u(1). \tag{6.3}$$

The operator $K_\lambda u := u(0) + Pu + \lambda Tu$ is compact and $K_0 u = u(0) + Pu \in \mathbb{R}^n$ for all u. According to the definition of the Leray–Schauder degree, if K_0 has no fixed points over the boundary of an open bounded set $\Omega \subset E$, then

$$deg(I - K_0, \Omega, 0) = deg((I - K_0)|_{\mathbb{R}^n}, \Omega \cap \mathbb{R}^n, 0).$$

This makes us particularly happy since for a constant function u it is clear that $u(0) = u$, so $I - K_0$ restricted to $\mathbb{R}^n$ is just the function $-P|_{\mathbb{R}^n}$. Moreover, from Exercise 5.2 we know that the degree of the latter function is equal, up to a sign, to that of $P|_{\mathbb{R}^n}$. Thus, we have proven the following theorem.

Theorem 6.2. *Assume there exists an open bounded $\Omega \subset C([0,1])$ such that*

1. Problem (6.3) has no solutions on $\partial \Omega$ for $0 < \lambda < 1$;
2. $\phi(u) \neq 0$ for $u \in \partial \Omega \cap \mathbb{R}^n$, where $\phi : \mathbb{R}^n \to \mathbb{R}^n$ is given by $\phi(u) := \int_0^1 f(t,u)\, dt$;
3. $deg(\phi, \Omega \cap \mathbb{R}^n, 0) \neq 0$.

Then (6.1) has at least one solution $u \in \overline{\Omega}$.

As an immediate application, we might try to see how the theorem works under condition (5.15) or, more generally,

$$f(t,u) \cdot \nabla G(u) \geq 0 \qquad \text{for} \qquad |u| \geq R, \tag{6.4}$$

where $G : \mathbb{R}^n \to \mathbb{R}$ is a C^1 function such that $G(u) \to +\infty$ as $|u| \to \infty$ and $\nabla G(u) \neq 0$ for $|u| \geq R$. The case (5.15) corresponds to $G(u) = |u|^2$; condition (6.4) looks more restrictive, although it shall be clear from the proof that it suffices to assume that (6.4) holds only for $|u| = r$ for some appropriate value of r.

Let us assume firstly that the inequality in (6.4) is strict. From Exercise 5.10, there exists $\tilde{r} \geq R$ such that $deg(\nabla G, B_r(0), 0) = 1$ when $r \geq \tilde{r}$. Thus, we may set $M := \max_{|u| \leq \tilde{r}} G(u)$ and r large enough such that $G(u) > M$ for $|u| \geq r$. Hence, if

$$\Omega := \{u \in C([0,1]) : \|G \circ u\|_\infty < M\},$$

then $B_R(0) \subset \Omega \cap \mathbb{R}^n \subset B_r(0)$ and, by the excision property, $deg(\nabla G, \Omega \cap \mathbb{R}^n, 0) = deg(\nabla G, B_r(0), 0) = 1$. Moreover, the homotopy $h(u, \lambda) := \lambda \phi(u) + (1 - \lambda)\nabla G(u)$ does not vanish over $\partial B_r(0)$, so $deg(\phi, \Omega \cap \mathbb{R}^n, 0) = 1$. Thus, we only need to prove that (6.3) with $0 < \lambda < 1$ has no solutions u with $\|G \circ u\|_\infty = M$.

Suppose that u is a solution such that $\|G \circ u\|_\infty = M$, and fix $t_0 \in [0,1)$ such that $|G(u(t_0))| = M$. Multiplying the equation by $\nabla G(u(t))$ we deduce that

$$0 \geq (G \circ u)'(t_0) = \lambda f(t, u(t_0)) \cdot \nabla G(u(t_0)) > 0,$$

a contradiction.

The general case follows from an approximation argument. From the strict case, the problem

$$u'(t) = f(t, u(t)) + \frac{1}{n} \nabla G(u(t)), \qquad u(0) = u(1)$$

has a solution u_n such that $\|u_n\|_\infty < r$. Moreover,

$$\|u_n'\|_\infty \leq \max_{0 \leq t \leq 1, |u| \leq r} |f(t,u)| + \max_{|u| \leq r} |\nabla G(u)|,$$

so by the Arzelá–Ascoli theorem we may extract a subsequence that converges to some function u. It is an easy exercise to prove that u is a solution of (6.1).

Remark 6.1. The reader may attempt a direct argument like that in Chap. 5: consider the family of problems

$$u'(t) = \lambda f(t, u(t)) + (1 - \lambda)\nabla F(u(t)), \qquad u(0) = u(1),$$

and prove, as before, that solutions are bounded. The function F_λ is now defined by $F_\lambda u := u - (u(0) + P_\lambda u + T_\lambda u)$, where

$$T_\lambda u(t) = \int_0^t \lambda f(s, u(s)) + (1 - \lambda)\nabla F(u(s))\, ds, \qquad P_\lambda u := T_\lambda u(1).$$

This setting has the advantage that no approximation argument is used, although the computation of $deg(F_0, \Omega, 0)$ is not direct and requires a new homotopy.

6.1.2 Better Late Than Never: Delayed Systems

Our second application concerns the existence of T-periodic solutions for a first-order delay differential system

$$u'(t) = f(t, u(t), u(t - \tau)), \tag{6.5}$$

with $\tau > 0$.

Here, the equation depends not only on the value of the solution at t but also on its value at $t - \tau$; thus, we shall assume that $f : \mathbb{R} \times \mathbb{R}^{2n} \to \mathbb{R}^n$ is continuous and T-periodic in t, that is, $f(t + T, u, v) = f(t, u, v)$ for all $(t, u, v) \in \mathbb{R} \times \mathbb{R}^{2n}$. We are looking for a "truly" T-periodic solution of (6.5), that is, a function u satisfying the equation and such that $u(t + T) = u(t)$ for all t. An appropriate Banach space for this problem is

$$C_T := \{u \in C(\mathbb{R}, \mathbb{R}^n) : u(t + T) = u(t) \text{ for all } t\}$$

equipped with the usual uniform norm. To obtain a continuation theorem for this case, let us firstly observe that the operator $u \mapsto u(0) + \int_0^t f(s, u(s), u(s - \tau))\, ds$, when applied to functions of C_T, does not necessarily return functions of C_T. Thus, a different formulation is needed: we cannot just copy what we did in the previous case. Again, the problem is resonant since the operator $L : D = C_T \cap C^1(\mathbb{R}, \mathbb{R}^n) \to C_T$ is noninvertible; as before, its kernel is the set of constants, identified with $\mathbb{R}^n$. Moreover, the range of L is the set of zero-average functions: indeed, if $u \in D$ and $u' = \varphi$, then $\overline{\varphi} := \frac{1}{T}\int_0^T \varphi(s)\, ds = 0$ and, conversely, for any $\varphi \in C_T$ such that $\overline{\varphi} = 0$ all its primitives $c + \int_0^t \varphi(s)\, ds$ belong to D, so $\varphi \in Im(L)$. Thus, it makes sense to define

$$\tilde{C}_T := \{\varphi \in C_T : \overline{\varphi} = 0\}$$

and a linear continuous right inverse of L, that is, $K : \tilde{C}_T \to D$, such that $LK\varphi = \varphi$ for all $\varphi \in \tilde{C}_T$. This can be done in infinitely many different ways since it depends on the choice of c as a function of φ. For convenience, from all possible primitives of φ we shall pick the (unique) one with zero average; in other words:

$$K\varphi(t) := -\frac{1}{T}\int_0^T \int_0^s \varphi(r)\, dr\, ds + \int_0^t \varphi(s)\, ds.$$

It is readily verified that $(K\varphi)' = \varphi$ and $K\varphi(t+T) - K\varphi(t) = \int_t^{t+T} \varphi(s)\,ds = 0$; moreover, $\int_0^T K\varphi(t)\,dt = 0$. Again, the proof of the compactness of K is left as an exercise.

Next, we define the operator $N : C_T \to C_T$ given by $Nu(t) := f(t, u(t), u(t-\tau))$. According to this setting, $u \in C_T$ is a solution if and only if two things happen:

$$\overline{Nu} = 0$$

and

$$u = \overline{u} + KNu.$$

It is important to notice that the first equality is a requirement for the second one, in the sense that otherwise Nu would not belong to the domain of K. This is not good if we want to write our problem as a fixed point equation since we should restrict ourselves to working in the set of functions satisfying the first equation. However, there is a simple way to avoid this problem: since we are pretending that both equations are satisfied simultaneously, it is not a big deal to add a zero term to the second one. In other words, the preceding system is equivalent to

$$\overline{Nu} = 0$$

and

$$u = \overline{u} + K(Nu - \overline{Nu}).$$

Finally, recall that the range of K is the set of zero-average functions of D; thus, a new trick will convert the system of two equations into a single one:

$$u = \overline{u} + \overline{Nu} + K(Nu - \overline{Nu}). \tag{6.6}$$

Indeed, if $u \in C_T$ is a solution of the original problem, then it clearly satisfies (6.6). Conversely, if $u \in C_T$ satisfies (6.6), then taking the average at both sides it follows that $\overline{Nu} = 0$; hence $u = \overline{u} + KNu$, which implies $u' = Nu$. With this in mind, we shall introduce the homotopy $h : C_T \times [0, 1] \to C_T$ given by

$$h(u, \lambda) = u - \left(\overline{u} + \overline{Nu} + \lambda K(Nu - \overline{Nu})\right). \tag{6.7}$$

When $\lambda > 0$, it is seen as before that $h(u, \lambda) = 0$ if and only if

$$u'(t) = \lambda f(t, u(t), u(t-\tau)). \tag{6.8}$$

Now suppose that h does not vanish when $u \in \partial\Omega$ for some $\Omega \subset C_T$ open and bounded. Then the degree of $h(\cdot, \lambda)$ at 0 over Ω is defined. Moreover, observe that $h(u, 0) - u = -(\overline{u} + \overline{Nu}) \in \mathbb{R}^n$, so we obtain

$$deg(h(\cdot, 0), \Omega, 0) = (-1)^n deg(\phi, \Omega \cap \mathbb{R}^n, 0),$$

with $\phi : \mathbb{R}^n \to \mathbb{R}^n$ given by $\phi(u) := -\overline{Nu} = \frac{1}{T} \int_0^T f(t, u, u)\,dt$. Thus, the following continuation theorem is proven.

Theorem 6.3. *Assume there exists an open bounded $\Omega \subset C_T$ such that*

1. Problem (6.8) has no solutions on $\partial\Omega$ for $0 < \lambda < 1$;
2. $\phi(u) \neq 0$ for $u \in \partial\Omega \cap \mathbb{R}^n$, with $\phi(u) := \frac{1}{T} \int_0^T f(t,u,u)\,dt$;
3. $\deg(\phi, \Omega \cap \mathbb{R}^n, 0) \neq 0$.

Then (6.5) has at least one solution $u \in \overline{\Omega}$.

Let us illustrate this latter result with a very concrete application for the Mackey–Glass equation

$$x'(t) = -a(t)x(t) + \frac{b(t)x(t)}{c(t) + x(t-\tau)^m} \tag{6.9}$$

modeling some physiological processes. Here, a, b, and c are continuous T-periodic functions, and we consider the problem of finding T-periodic positive solutions of (6.9). Substitution of $u(t) = \ln x(t)$ yields the following problem for u:

$$u'(t) = -a(t) + \frac{b(t)}{c(t) + e^{mu(t-\tau)}} := f(t, u(t-\tau)).$$

Here,

$$\phi(u) = -\overline{a} + \frac{1}{T} \int_0^T \frac{b(t)}{c(t) + e^{mu}}\,dt,$$

so it is easily verified that ϕ is strictly nonincreasing and

$$\lim_{u \to -\infty} \phi(u) = -\overline{a} + \overline{b/c}, \qquad \lim_{u \to +\infty} \phi(u) = -\overline{a}.$$

Thus, ϕ has a (unique) zero if and only if $\overline{b/c} > \overline{a}$; in this case, the degree of ϕ over any open interval containing its zero is -1.

We claim that the condition $\overline{b/c} > \overline{a}$ is sufficient for the existence of solutions; moreover, it shall follow from the proof that the condition is also necessary. Indeed, setting

$$\Omega := \{u \in C_T : \alpha < u(t) < \beta \text{ for all } t\},$$

for appropriate $\alpha < \phi^{-1}(0) < \beta$, it suffices to show that the problem

$$u'(t) = \lambda f(t, u(t-\tau)) \tag{6.10}$$

has no solutions on $\partial\Omega$ for $0 < \lambda < 1$. Assume that $u \in C_T$ solves (6.10), take the average, and divide by λ to deduce that

$$0 = -\overline{a} + \frac{1}{T} \int_0^T \frac{b(t)}{c(t) + e^{mu(t-\tau)}}\,dt \geq -\overline{a} + \frac{1}{T} \int_0^T \frac{b(t)}{c(t) + e^{mu_{max}}}\,dt = \phi(u_{max}),$$

where u_{max} denotes the maximum value of u. In the same way, it is seen that $\phi(u_{min}) \geq 0$, and since $\phi^{-1} : (-\overline{a}, -\overline{a} + \overline{b/c}) \to \mathbb{R}$ is also nonincreasing we obtain

$$u_{min} \leq \phi^{-1}(0) \leq u_{max}.$$

In particular, there exists t_0 such that $u(t_0) = \phi^{-1}(0)$. Furthermore, it follows from (6.10) that

$$|u'(t)| < a(t) + \frac{b(t)}{c(t) + e^{mu(t-\tau)}},$$

and thus

$$\frac{1}{T}\int_0^T |u'(t)|\,dt < \bar{a} + \frac{1}{T}\int_0^T \frac{b(t)}{c(t) + e^{mu(t-\tau)}}\,dt = 2\bar{a}.$$

Writing $u(t) = u(t_0) + \int_{t_0}^t u'(s)\,ds$ we conclude that

$$\phi^{-1}(0) - 2T\bar{a} < u(t) < \phi^{-1}(0) + 2T\bar{a}$$

for all t. Hence, it suffices to take

$$\alpha := \phi^{-1}(0) - 2T\bar{a}, \qquad \beta := \phi^{-1}(0) + 2T\bar{a},$$

since, by monotonicity, $\phi(\alpha) > 0 > \phi(\beta)$, and hence $deg(\phi,(\alpha,\beta),0) = -1$.

Also, we may solve a matter that has left readers *in suspense* since Chap. 4: the problem

$$x'(t) = -a(t)x(t) + sb(t)g(x(t - \tau(t))), \tag{6.11}$$

with $a, b, \tau : \mathbb{R} \to (0, +\infty)$ continuous and T-periodic, $s > 0$, and $g : [0, +\infty) \to (0, +\infty)$ continuous, nondecreasing, and superlinear. We have shown that, for some $s^* > 0$, the problem has no T-periodic positive solutions when $s > s^*$ and at least one when $s < s^*$. Using the previous continuation theorem, we shall prove that, if s is small enough, then the problem has at least two positive T-periodic solutions. Indeed, observe in the first place that $\phi(x) = -\bar{a}x + s\overline{g(x)b}$, and, hence, for any fixed $s > 0$,

$$\lim_{x \to 0^+} \frac{\phi(x)}{x} = \lim_{x \to +\infty} \frac{\phi(x)}{x} = +\infty.$$

Thus, $\phi(x) > 0$ for $x \gg 0$ or $x \ll 1$; in particular, the set

$$\Omega_{r,R} := \{x \in C_T : r < x(t) < R \text{ for all } t\}$$

is not a good choice when $r > 0$ is very small and $R > 0$ is very large because in that case $deg(\phi,(r,R),0) = 0$. Nevertheless, if there exists some M between r and R such that $\phi(M) < 0$, then $deg(\phi,(r,M),0) = -1 = -deg(\phi,(M,R),0)$, so there might be some hope of finding a solution in $\Omega_{r,M}$ and another one in $\Omega_{M,R}$.

To this end, observe that if $x \in C_T$ is positive and satisfies

$$x'(t) = \lambda(-a(t)x(t) + sb(t)g(x(t - \tau(t)))) \tag{6.12}$$

for some $\lambda \in (0,1)$, then $\frac{x'(t)}{x(t)} > -a(t)$, and integrating we deduce that

$$\log(x(t_1)) - \log(x(t_0)) > -\int_{t_0}^{t_1} a(t)\,dt$$

for arbitrary $t_0 < t_1$. In particular, from the periodicity of x we deduce that

$$\log(x_{min}) - \log(x_{max}) > -\int_0^T a(t)\,dt,$$

that is,

$$x_{min} \geq Cx_{max},$$

with $C = e^{-\int_0^T a(t)\,dt}$.

Fix $t_0 \in [0,T]$ such that $x(t_0) = x_{max}$; then from (6.12) we obtain

$$a(t_0)x(t_0) = sb(t_0)g(x(t_0 - \tau(t_0))),$$

and thus

$$\frac{a(t_0)}{sb(t_0)} = \frac{g(x(t_0 - \tau(t_0)))}{x(t_0 - \tau(t_0))}\frac{x(t_0 - \tau(t_0))}{x_{max}} \geq C\frac{g(x(t_0 - \tau(t_0)))}{x(t_0 - \tau(t_0))}.$$

Next, fix $R > 0$ such that $C\frac{g(x)}{x} > \left(\frac{a}{sb}\right)_{max}$ when $x \geq CR$. If $x_{max} \geq R$, then $x(t) \geq CR$ for all t, and from the previous inequality $\frac{a(t_0)}{sb(t_0)} > \left(\frac{a}{sb}\right)_{max}$, a contradiction. In the same way, if $t_0 \in [0,T]$ is now chosen in such a way that $x(t_0) = x_{min}$, then

$$\frac{a(t_0)}{sb(t_0)} = \frac{g(x(t_0 - \tau(t_0)))}{x(t_0 - \tau(t_0))}\frac{x(t_0 - \tau(t_0))}{x_{min}} \geq C\frac{g(x(t_0 - \tau(t_0)))}{x(t_0 - \tau(t_0))},$$

so fixing $r > 0$ such that $C\frac{g(x)}{x} > \left(\frac{a}{sb}\right)_{max}$ when $x \leq \frac{r}{C}$ we also deduce that $x_{min} > r$.

Note that the previous choice of r and R depends only on s; thus, our strategy shall consist in fixing a value $M > 0$ and finding $s_* > 0$ such that (6.12) has no positive solutions with $x_{max} = M$ or $x_{min} = M$ and $\phi(M) < 0$ when $s < s_*$. Thus, for each $s < s_*$ one obtains $r < M < R$ as before, and the existence of at least two solutions is guaranteed.

Fix an arbitrary $M > 0$, and let

$$\gamma := \inf_{M \leq x \leq \frac{M}{C}} \frac{x}{g(x)}.$$

Assume that $x \in C_T$ satisfies (6.12) with $x_{max} = M$, and let t_0 be a value where x_{max} is achieved. Then

$$a(t_0)M = sb(t_0)g(t_0 - x(t_0 - \tau(t_0))) \leq sb(t_0)g(M),$$

and we deduce that $s \geq \gamma\left(\frac{a}{b}\right)_{min}$. In the same way, if we suppose now that $x(t_0) = x_{min} = M$ and $x(t_0) = x_{min}$, then

$$s = \frac{a(t_0)M}{b(t_0)g(t_0 - x(t_0 - \tau(t_0)))} = \frac{a(t_0)x(t_0 - \tau(t_0))}{b(t_0)g(t_0 - x(t_0 - \tau(t_0)))}\frac{M}{x(t_0 - \tau(t_0))} \geq \frac{a(t_0)}{b(t_0)}C\gamma,$$

so we conclude that $s \geq C\gamma \left(\frac{a}{b}\right)_{min}$. Summarizing, if we fix $s < s_* := C\gamma \left(\frac{a}{b}\right)_{min}$ and $r < M < R$ as before, then (6.12) has no solutions on $\partial \Omega_{r,M} \cup \partial \Omega_{M,R}$. Moreover, observe that

$$\phi(M) = M\left(-\overline{a} + s\overline{b}\frac{g(M)}{M}\right) \leq M\left(-\overline{a} + \frac{s\overline{b}}{\gamma}\right) < M\left(-\overline{a} + \overline{b}C\left(\frac{a}{b}\right)_{min}\right).$$

Using the Cauchy mean value theorem, it is seen that $\left(\frac{a}{b}\right)_{min} \leq \frac{\overline{a}}{\overline{b}}$; furthermore, since $C < 1$, we conclude that $\phi(M) < 0$.

6.1.3 Periodic Solutions of Second-Order Equations: Landesman, Lazer, Leach... and Many More

As a third application of the continuation method, consider the second-order problem

$$u''(t) = f(t, u(t), u'(t)) \tag{6.13}$$

under periodic conditions

$$u(0) = u(1), \qquad u'(0) = u'(1). \tag{6.14}$$

We shall assume that $f : [0,1] \times \mathbb{R}^{2n} \to \times \mathbb{R}^n$ is continuous. Due to the dependence on u', now we may take $F = C([0,1])$ as before, but $E := C^1([0,1])$. The operator $Lu := u''$ can be defined in $D := \{u \in C^2([0,1]) : u(0) = u(1), u'(0) = u'(1)\}$; as in the previous cases, it is verified that

$$Ker(L) = \mathbb{R}^n, \qquad Im(L) = \{\varphi \in C([0,1]) : \overline{\varphi} = 0\}.$$

Again, we may construct a (compact) right inverse $K : Im(L) \to D$ of the operator L by considering $K\varphi = u$, the unique solution $u \in D$ of the problem $u'' = \varphi$, such that $\overline{u} = 0$.

Let $N : E \to F$ be given by $Nu(t) = f(t, u(t), u'(t))$; then a function $u \in D$ satisfies the equation $Lu = Nu$ if and only if

$$u - \overline{u} = KNu, \qquad \overline{Nu} = 0.$$

Indeed, if $Lu = Nu$, then integration at both sides yields $\overline{Nu} = 0$; moreover, as $LKNu = Nu$ we deduce that u and KNu differ in a constant and, taking the average, it is seen that $u = \overline{u} + KNu$. The converse follows from direct differentiation of the equality $u - \overline{u} = KNu$. As in the previous case, we arrive at a system in which the first equation does not make sense if the second one is not satisfied; as seen, we can make both equations independent of each other simply by considering the equivalent system

$$u - \overline{u} = K(Nu - \overline{Nu}), \qquad \overline{Nu} = 0,$$

or, even better, the single equation

$$u = \bar{u} + \overline{Nu} + K(Nu - \overline{Nu}).$$

Thus, we may use once again the homotopy given by (6.7), now defined from $C^1([0,1]) \times [0,1]$ to $C([0,1])$. When $\lambda > 0$, it is seen that $h(u,\lambda) = 0$ if and only if u is a solution of

$$u''(t) = \lambda f(t, u(t), u'(t)), \qquad u(0) = u(1), \ u'(0) = u'(1), \qquad (6.15)$$

so the following continuation theorem is obtained.

Theorem 6.4. *Assume there exists an open bounded $\Omega \subset C^1([0,1])$ such that*

1. Problem (6.15) has no solutions on $\partial\Omega$ for $0 < \lambda < 1$;
2. $\phi(u) \neq 0$ for $u \in \partial\Omega \cap \mathbb{R}^n$, with $\phi(u) := \int_0^1 f(t,u,0)\,dt$;
3. $deg(\phi, \Omega \cap \mathbb{R}^n, 0) \neq 0$.

Then (6.13)–(6.14) has at least one solution $u \in \overline{\Omega}$.

It is worth noting that the presence of a mysterious 0 in the definition of ϕ has a simple explanation. As in the previous cases, the degree of $h(\cdot,0)$ is defined as the degree of its restriction to $\Omega \cap \mathbb{R}^n$, and in that subset, the derivative of any u is equal to 0. This is certainly a trivial comment, but it should keep us alert regarding other cases in which $Ker(L)$ is more complicated. Also, if f does not depend on u', then we may take Ω as a bounded open subset of $C([0,1])$ instead of $C^1([0,1])$.

To warm up, we may start with an elementary application to a case already studied in Chap. 1 by the shooting method and revisited in Chaps. 4 and 5: the (strict) Hartman condition

$$f(t,u) \cdot u > 0 \qquad \text{for } |u| = R. \qquad (6.16)$$

As shown in the previous chapter, a direct degree argument makes it possible to prove the existence of solutions for periodic conditions; however, since the problem is resonant, it is not a bad idea to check how it can be solved in light of Theorem 6.4. To this end, just take $\Omega := \{u \in C([0,1]) : \|u\|_\infty < R\}$. The fact that (6.15) has no solutions on $\partial\Omega$ for $0 < \lambda < 1$ is now familiar to us: if a solution u touches the boundary of $B_R(0) \subset \mathbb{R}^n$ at some t_0 from inside, then the second derivative of the function $|u(t)|^2$ at t_0 must be nonpositive, and a contradiction is obtained. Further, condition (6.16) implies that ϕ is homotopic to the identity over $B_R(0)$, so $deg(\phi, \Omega \cap \mathbb{R}^n, 0) = 1$. Certainly, the general results proven in [45] are far from being so easy. Something else about this condition shall be said in Sect. 6.4.

We may illustrate further uses of Theorem 6.4 with a famous example: the Landesman–Lazer conditions, first obtained in [67] for a resonant elliptic second-order scalar equation under Dirichlet conditions (for a survey on Landesman–Lazer conditions see, e.g., [73]). Adapted to the problem

$$u''(t) + g(u(t)) = p(t), \qquad u(0) = u(1), \quad u'(0) = u'(1) \qquad (6.17)$$

the Landesman–Lazer theorem says, roughly, that if the average of the forcing term p lies between the limits at $\pm\infty$ of the bounded and continuous nonlinearity $g : \mathbb{R} \to \mathbb{R}$, then the problem admits at least one solution. Furthermore, if $g(u) \neq g(\pm\infty)$ for all u, then the conditions are also necessary. The latter assertion is easily verified since integration of (6.17) yields, by periodicity,

$$\int_0^1 g(u(t))\, dt = \overline{p},$$

and hence, from the fact that the range of the function u is compact, we obtain:

$$\min\{g(-\infty), g(+\infty)\} < \min_{t \in [0,1]} g(u(t)) \leq \overline{p} \leq \max_{t \in [0,1]} g(u(t)) < \max\{g(-\infty), g(+\infty)\}.$$

Thus, we may focus on the existence part, namely:

Theorem 6.5. *Let $g : \mathbb{R} \to \mathbb{R}$ be continuous and bounded, and assume that the limits*

$$g(\pm\infty) = \lim_{u \to \pm\infty} g(u)$$

exist. Then (6.17) *has at least one solution for any $p \in C([0,1])$ such that*

$$g(-\infty) < \overline{p} < g(+\infty), \tag{6.18}$$

$$g(+\infty) < \overline{p} < g(-\infty). \tag{6.19}$$

Observe that when a constant is added to g, it may always be assumed, without loss of generality, that $\overline{p} = 0$. The theorem requires that $g(-\infty) \neq g(+\infty)$, and if this happens, then obviously just one of the assumptions (6.18) and (6.19) is possible. Nevertheless, it is worth writing both conditions separately to emphasize the two different situations. Note, for instance, that the problem of proving the existence of solutions under the second condition has been solved (or at least suggested) under weaker assumptions in Exercise 5.21; in particular, the limits may not exist. Later on, we shall discuss extensions of Theorem 6.5 in different directions; for the moment, it is worth giving a short proof using Theorem 6.4. To this end, it suffices to consider

$$\Omega = \{u \in C([0,1]) : \|u\|_\infty < R\}$$

and prove that, for sufficiently large R:

1. If
$$u''(t) = \lambda[p(t) - g(u(t))], \qquad u(0) = u(1), \quad u'(0) = u'(1) \tag{6.20}$$

   for some $\lambda \in (0,1)$, then $\|u\|_\infty < R$;
2. $\phi(-R)\phi(R) < 0$, where $\phi : \mathbb{R} \to \mathbb{R}$ is given by $\phi(u) = \overline{p} - g(u)$.

In other words, the original problem is reduced to the basic Bolzano rule, which states that if a real function has different signs at the ends of an interval, then its degree is ± 1. In this sense, the method is very powerful, although it has the nontrivial requirement, expressed in the first condition, that the homotopy does not vanish over the boundary of Ω.

By contradiction, suppose that u_n is a solution of (6.20) for some $\lambda_n \in (0,1)$ such that $\|u_n\|_\infty \to \infty$. Then $\|u_n''\|_\infty \leq \|p\|_\infty + \|g\|_\infty$, and, as in previous chapters, we deduce

$$\|u_n - \overline{u_n}\|_\infty \leq \frac{1}{2}\|u_n'\|_\infty \leq \frac{1}{4\pi}\|u_n''\|_\infty < \frac{\|p\|_\infty + \|g\|_\infty}{4\pi}.$$

Thus, writing $\|u_n\|_\infty \leq \|u_n - \overline{u_n}\|_\infty + |\overline{u_n}|$ it is seen that $|\overline{u_n}| \to \infty$. Taking a subsequence, we may suppose, for example, that $\overline{u_n} \to +\infty$; in this case, writing $u_n(t) = \overline{u_n} + u_n(t) - \overline{u_n}$ we deduce, moreover, that $u_n(t) \to +\infty$ uniformly in t.

On the other hand, integration of (6.20) yields, as $\lambda_n \neq 0$,

$$\int_0^1 g(u_n(t))\,dt = \overline{p},$$

and letting $n \to \infty$ we deduce $\overline{p} = g(+\infty)$, a contradiction. The proof is analogous if $\overline{u_n} \to -\infty$. Finally, observe that both (6.18) and (6.19) imply that $\phi(-R)$ and $\phi(R)$ have different signs when R is large, so the proof is complete.

Remark 6.2. Using critical point theory (e.g., [77]), it is not difficult to prove the existence of solutions when

$$\limsup_{u \to -\infty} \frac{G(u)}{u} < \overline{p} < \liminf_{u \to +\infty} \frac{G(u)}{u}$$

or

$$\limsup_{u \to +\infty} \frac{G(u)}{u} < \overline{p} < \liminf_{u \to -\infty} \frac{G(u)}{u},$$

where $G(u) := \int_0^u g(s)\,ds$. These conditions are sometimes referred to as *potential Landesman–Lazer conditions* and are weaker than the original ones. Indeed, if, for example, $g(+\infty) > \overline{p}$, then fix $\varepsilon > 0$ small and u_0 such that $g(u) > \overline{p} + \varepsilon$ for all $u > u_0$. Then

$$G(u) = G(u_0) + \int_{u_0}^u g(s)\,ds > G(u_0) + (\overline{p} + \varepsilon)(u - u_0),$$

and hence $\liminf_{u \to +\infty} \frac{G(u)}{u} \geq \overline{p} + \varepsilon$. The potential Landesman–Lazer conditions take an even weaker form in the main result of [2] that, adapted to our situation, ensures the existence of solutions if

$$\lim_{|u| \to \infty} [G(u) - \overline{p}u] = +\infty$$

or

$$\lim_{|u| \to \infty} [G(u) - \overline{p}u] = -\infty.$$

Note that, in contrast with the previous assumptions, it might happen that $\frac{G(u)}{u} \to \overline{p}$, although not too fast. As mentioned, we shall introduce more generalizations that in some cases do not prescribe any asymptotic behavior at all, either for g or its primitive. We also remark that critical point methods cannot be directly applied if a term $au'(t)$ with $a \neq 0$ is added to the equation. This justifies the use of topological methods.

For a better understanding of the Landesman–Lazer theorem, observe that the average $\overline{p}$ can be regarded as the projection (in the L^2 sense) of p into the kernel of the linear operator $Lu := u''$, which, as mentioned, is the set of constant functions, identified with $\mathbb{R}$. The original result by Landesman and Lazer was established for a more general situation, but still with a one-dimensional kernel. To fix ideas, we may suppose now that the equation is

$$Lu(t) + g(u(t)) = p(t),$$

with $L : D \subset C([0,1]) \to C([0,1])$ symmetric (or "formally self-adjoint") with respect to the inner product of $L^2(0,1)$ and $Ker(L) = \langle \varphi \rangle := span\{\varphi\}$ for some $\varphi \in D$. Symmetry of L implies that $Ker(L)$ is orthogonal to $Im(L)$ since $\int_0^1 \varphi(t)L\psi(t)\,dt = \int_0^1 L\varphi(t)\psi(t)\,dt = 0$ for all ψ. Also, note that $C([0,1])$ can be written as a direct sum $\langle \varphi \rangle \oplus \langle \varphi \rangle^\perp$, where $\langle \varphi \rangle^\perp$ is the "orthogonal complement" defined by

$$\langle \varphi \rangle^\perp := \{ \psi \in C([0,1]) : \int_0^1 \psi(t)\varphi(t)\,dt = 0 \}.$$

If, for simplicity, we assume $\|\varphi\|_{L^2} = 1$, then any $u \in C([0,1])$ is uniquely decomposed as $u = Pu + (u - Pu)$, where

$$Pu = \left(\int_0^1 u(t)\varphi(t)\,dt \right) \varphi.$$

Now suppose that $Im(L) = \langle \varphi \rangle^\perp$ and, furthermore, that L has a compact inverse $K : Im(L) \to \langle \varphi \rangle^\perp \cap D$. In this case, the continuity of K implies the existence of a constant c such that $\|K\psi\|_\infty \leq c\|\psi\|_\infty$ for all $\psi \in Im(L)$ or, equivalently, $\|u - Pu\|_\infty \leq c\|Lu\|_\infty$ for all $u \in D$.

Equipped with this machinery, it is not hard to conceive a somewhat more abstract version of the Landesman–Lazer theorem, after introducing some minor changes to Theorem 6.4. To this end, observe that the homotopy equation $Lu = \lambda(p - g(u))$ with $\lambda > 0$ yields

$$\|u - Pu\|_\infty \leq c\|Lu\|_\infty \leq c(\|p\|_\infty + \|g\|_\infty)$$

and

$$\int_0^1 g(u(t))\varphi(t)\,dt = \int_0^1 p(t)\varphi(t)\,dt.$$

In particular, if we suppose as before that u_n is a solution for some $\lambda_n \in (0,1)$ with $\|u_n\|_\infty \to \infty$, then $\|u_n - Pu_n\|_\infty$ is bounded, and thus $\|Pu_n\|_\infty \to \infty$. We may write

$Pu_n = r_n \varphi$ and, passing to a subsequence, assume, for example, that $r_n \to +\infty$. The difference with the preceding case is that, here, the function φ may change sign, so now $u_n(t) = u_n(t) - Pu_n(t) + r_n \varphi(t)$ tends to $+\infty$ when $\varphi(t) > 0$ and to $-\infty$ when $\varphi(t) < 0$. Thus, the integral in the previous inequality must be split into two parts:

$$\int_0^1 p(t)\varphi(t)\,dt = \int_{\{\varphi>0\}} g(u_n(t))\varphi(t)\,dt + \int_{\{\varphi<0\}} g(u_n(t))\varphi(t)\,dt.$$

Taking limit, we arrive at the equality

$$\int_0^1 p(t)\varphi(t)\,dt = g(+\infty)\int_{\{\varphi>0\}} \varphi(t)\,dt + g(-\infty)\int_{\{\varphi<0\}} \varphi(t)\,dt$$

or

$$\int_0^1 p(t)\varphi(t)\,dt = g(-\infty)\int_{\{\varphi>0\}} \varphi(t)\,dt + g(+\infty)\int_{\{\varphi<0\}} \varphi(t)\,dt$$

if we suppose, instead, $r_n \to -\infty$. This explains the standard form of the remarkable Landesman–Lazer conditions:

$$g(-\infty)\int_{\{\varphi>0\}} \varphi(t)\,dt + g(+\infty)\int_{\{\varphi<0\}} \varphi(t)\,dt < \langle p, \varphi \rangle$$

$$< g(+\infty)\int_{\{\varphi>0\}} \varphi(t)\,dt + g(-\infty)\int_{\{\varphi<0\}} \varphi(t)\,dt$$

or vice versa. Observe that, in this case, the analogue to the function ϕ in Theorem 6.4 is defined from the subspace $\langle \varphi \rangle$ into itself as

$$\phi(u) = P(p - g(u)) = \left(\int_0^1 p(t)\varphi(t)\,dt - \int_0^1 g(u(t))\varphi(t)\,dt \right)\varphi.$$

But we are only interested in computing its degree, so, writing $u = r\varphi$ with $r \in \mathbb{R}$, we may just regard ϕ as the real function

$$\phi(r) = \int_0^1 p(t)\varphi(t)\,dt - \int_0^1 g(r\varphi(t))\varphi(t)\,dt.$$

Thus, for R large enough, the previous conditions guarantee that $\phi(R)$ and $\phi(-R)$ have different signs.

To give a concrete example, we may, for convenience, rescale the interval and consider the Dirichlet problem

$$u''(t) + n^2 u(t) + g(u(t)) = p(t), \qquad u(0) = u(\pi) = 0.$$

For arbitrary $n \in \mathbb{N}$, the kernel of the operator $L_n u := u'' + n^2 u$ is the subspace spanned by the function $\varphi_n(t) := \sin(nt)$. The reader may verify that $Im(L_n) = \langle \varphi_n \rangle^\perp$ by direct computation or, more elegantly, using standard arguments of functional analysis (also, Exercise 5.18 can be applied). When $n = 1$, the function φ_n

does not change sign and the Landesman–Lazer conditions simply read

$$2g(-\infty) < \int_0^\pi p(t)\sin(t)\,dt < 2g(+\infty)$$

or

$$2g(+\infty) < \int_0^\pi p(t)\sin(t)\,dt < 2g(-\infty).$$

Theorem 6.5 admits an even more abstract formulation; besides the proof given in the preceding paragraphs, some other nice and simple proofs like the one in [51] can be easily adapted to very general contexts.

A more difficult situation occurs when the kernel of the linear operator has a dimension higher than 1; for example, the problem

$$u'' + n^2 u + g(u) = p(t), \qquad u(0) = u(2\pi), \qquad u'(0) = u'(2\pi), \qquad (6.21)$$

with n an integer. When $n = 0$, we are back at the previous situation with $Ker(L) = \mathbb{R}$; for $n \in \mathbb{N}$ the situation is usually referred to in the literature as *resonance at a higher-order eigenvalue*. This name has its origins in the eigenvalue problem

$$-u'' = \lambda u, \qquad u(0) = u(2\pi), \qquad u'(0) = u'(2\pi),$$

which admits nontrivial solutions if and only if $\lambda \in \mathbb{N}_0^2$. For $\lambda = n^2 > 0$, the set of eigenfunctions is the subspace of $C^2([0, 2\pi])$ spanned by $\sin nt$ and $\cos nt$; in other words, if we set again $L_n u = u'' + n^2 u$, now thought of as an operator from the space $D := \{u \in C^2([0, 2\pi]) : u(0) = u(2\pi), u'(0) = u'(2\pi)\}$ into $C([0, 2\pi])$, then

$$ker(L_n) = \{\alpha \cos nt + \beta \sin nt : \alpha, \beta \in \mathbb{R}\} := V_n.$$

Recall that when $n = 0$, the Landesman–Lazer conditions impose that $\overline{p}$, the projection of p into $ker(L_0)$, must lie between the limits of g at $\pm\infty$; thus, we might expect an analogous condition for $n > 0$, now expressed in terms of the nth Fourier coefficients of p, that is,

$$\alpha_n(p) = \frac{1}{\pi} \int_0^{2\pi} p(t)\cos nt\,dt, \qquad \beta_n(p) = \frac{1}{\pi} \int_0^{2\pi} p(t)\sin nt\,dt.$$

In 1969, shortly before [67] was published, Lazer and Leach had shown that this was, indeed, the case.

Theorem 6.6 ([70]). *Assume that $g : \mathbb{R} \to \mathbb{R}$ is continuous and bounded and that the limits $g(\pm\infty)$ exist. If*

$$\sqrt{\alpha_n(p)^2 + \beta_n(p)^2} < \frac{2}{\pi}|g(+\infty) - g(-\infty)|, \qquad (6.22)$$

then (6.21) has at least one 2π-periodic solution.

In other words, (6.22) says that the projection of p into the kernel must be "small." As happens when $n = 0$, it is easy to prove that the condition is also necessary when $g(u)$ is strictly between $g(-\infty)$ and $g(+\infty)$ for all u. To prove this latter fact, complex notation is of great help: multiply the equation by e^{int} and integrate to obtain

$$\int_0^{2\pi} [u''(t) + n^2 u(t) + g(u(t))] e^{int}\, dt = \pi(\alpha_n(p) + i\beta_n(p))$$

$$\frac{1}{\pi} \int_0^{2\pi} g(u(t)) e^{int}\, dt = \rho e^{i\omega}$$

for some $\omega \in [0, 2\pi)$, where $\rho = \sqrt{\alpha_n(p)^2 + \beta_n(p)^2}$. Using periodicity,

$$\rho = \frac{1}{\pi} \int_0^{2\pi} g(u(t)) e^{i(nt-\omega)}\, dt = \frac{1}{\pi} \int_0^{2\pi} g(u(t + \tfrac{\omega}{n})) e^{int}\, dt,$$

so the imaginary part vanishes and

$$\rho = \frac{1}{\pi} \int_0^{2\pi} g(u(t + \tfrac{\omega}{n})) \cos nt\, dt.$$

Suppose, for example, $g(-\infty) < g(+\infty)$, and let $I_n^{\pm}$ be the subsets of $[0, 2\pi]$ in which $\cos nt > 0$ and $\cos nt < 0$, respectively. Because the image of u is compact, we deduce that

$$\int_{I_n^+} g(u(t + \omega)) \cos nt\, dt < \int_{I_n^+} g(+\infty) \cos nt\, dt = 2g(+\infty)$$

and

$$\int_{I_n^-} g(u(t + \omega)) \cos nt\, dt < \int_{I_n^-} g(-\infty) \cos nt\, dt = -2g(-\infty),$$

so $\rho < \frac{2}{\pi}(g(+\infty) - g(-\infty))$.

Now we turn to existence. Direct computation (or, as before, "elegant" arguments from functional analysis) show that if we regard L_n as an operator from D to $C([0, 2\pi])$, where $D = \{u \in C^2([0, 2\pi]) : u(0) = u(2\pi), u'(0) = u'(2\pi)\}$, then

$$\mathrm{Im}(L_n) = V_n^{\perp} := \left\{ \varphi \in C([0, 2\pi]) : \int_0^{2\pi} \varphi(t) \cos nt\, dt = \int_0^{2\pi} \varphi(t) \sin nt\, dt = 0 \right\}.$$

Next, consider the orthogonal (in the L^2 sense) projection P_n of $C([0, 2\pi])$ onto V_n given by $P_n u(t) = \alpha_n(u) \cos nt + \beta_n(u) \sin nt$ with α_n and β_n as before. There are several ways to prove the existence of a constant c such that $\|u - P_n u\|_\infty \leq c\|L_n u\|_\infty$ for all $u \in D$; for example, using the Fourier series expansion of u it is readily seen that

$$\|u - P_n u\|_{L^2}^2 + \|(u - P_n u)'\|_{L^2}^2 \leq \|L_n u\|_{L^2}^2,$$

from where the desired inequality is obtained. Finally, define the (compact) mapping $K_n : V_n^{\perp} \to V_n^{\perp}$ given by $K_n \varphi = u$, where u is the unique solution of

$$L_n u = \varphi, \qquad u(0) = u(2\pi), \; u'(0) = u'(2\pi), \; P_n u = 0.$$

In this setting, problem (6.21) is written equivalently as

$$u - P_n u = KNu, \qquad P_n Nu = 0,$$

where $Nu := p - g(u)$, or

$$u = P_n u + P_n Nu + K(Nu - P_n u).$$

So we define $F_\lambda : C([0, 2\pi]) \to C([0, 2\pi])$, as in the previous examples, given by $F_\lambda u := u - [P_n u + P_n Nu + \lambda K(Nu - P_n u)]$. Let us firstly prove that $F_\lambda(u) \neq 0$ for $\lambda \in (0, 1)$ and $\|u\|_\infty = R$ large enough. Indeed, if $F_{\lambda_k} u_k = 0$ for $\lambda_k \in (0, 1)$ and $\|u_k\|_\infty \to \infty$, then from the equality

$$u_k'' + n^2 u_k = \lambda_k(g(u_k) - p)$$

we deduce that $P_n(g(u_k) - p) = 0$ and

$$\|u_k - P_n(u_k)\|_\infty \leq c\|g(u_k) - p\|_\infty \leq c(\|g\|_\infty + \|p\|_\infty),$$

so $\|P_n u_k\|_\infty \to \infty$. Next, write $P_n u_k(t) = \rho_k \cos(nt - \omega_k)$, where $\rho_k \to +\infty$ and $\omega_k \in [0, 2\pi]$. We may assume that ω_k converges to some ω, and from the identity $P_n(g(u_k)) = P_n(p)$ we obtain (using complex notation again!)

$$\frac{1}{\pi} \int_0^{2\pi} g(u_k(t)) e^{int} \, dt = \alpha_n(p) + i\beta_n(p).$$

Now write $u_k(t) = A_k(t) + \rho_k \cos(nt - \omega_k)$ to deduce, by substitution,

$$\int_0^{2\pi} g(u_k(t)) e^{int} \, dt = e^{i\omega_k} \int_0^{2\pi} g(A_k(t + \tfrac{\omega_k}{n}) + \rho_k \cos nt)) e^{int} \, dt.$$

As before, it is seen that

$$\int_{I+} g(A_k(t) + \rho_k \cos(nt - \omega_k)) e^{int} \, dt \to g(+\infty) \int_{I+} e^{int} \, dt = 2g(+\infty)$$

and

$$\int_{I-} g(A_k(t) + \rho_k \cos(nt - \omega_k)) e^{int} \, dt \to g(-\infty) \int_{I-} e^{int} \, dt = -2g(-\infty)$$

since $\int_{I+} \sin nt \, dt = \int_{I-} \sin nt \, dt = 0$. Thus,

$$\alpha_n(p) + i\beta_n(p) = \frac{1}{\pi} \int_0^{2\pi} g(u_k(t)) e^{int} \, dt \to \frac{2}{\pi}(g(+\infty) - g(-\infty)) e^{i\omega},$$

which contradicts the fact that $|\alpha_n(p) + i\beta_n(p)| < \frac{2}{\pi}|g(+\infty) - g(-\infty)|$.

Now what about F_0? In this case, we only need to consider its restriction to V_n. For simplicity, we may ignore the isomorphism $\alpha + \beta i \mapsto \alpha \cos nt + \beta \sin nt$, so the mapping F_0 can be directly interpreted as a complex function:

$$F_0(\alpha + \beta i) = \frac{1}{\pi} \int_0^{2\pi} g(\alpha \cos nt + \beta \sin nt) e^{int} dt - (\alpha_n(p) + i\beta_n(p)).$$

Again, we may write $\alpha \cos nt + \beta \sin nt = \rho \cos(nt - \omega)$ and, as before,

$$\int_0^{2\pi} g(\rho \cos(nt - \omega)) e^{int} dt = e^{i\omega} \int_0^{2\pi} g(\rho \cos(nt)) e^{int} dt \to 2[g(+\infty) - g(-\infty)] e^{i\omega}$$

as $\rho \to +\infty$. Observe that the imaginary part of the integrand in the middle term is an odd function; thus, fixing ρ large enough we deduce that

$$F_0 u = C e^{i\omega} - (\alpha_n(p) + i\beta_n(p))$$

for some real constant C such that $|C| > |\alpha_n(p) + i\beta_n(p)|$. This shows that the restriction of F_0 to the boundary of a large disk completes one turn around $\alpha_n(p) + i\beta_n(p)$, so for R large, $\deg(F_0, B_R(0), 0) = \pm 1$, where the sign depends on the respective cases $C > 0$ and $C < 0$.

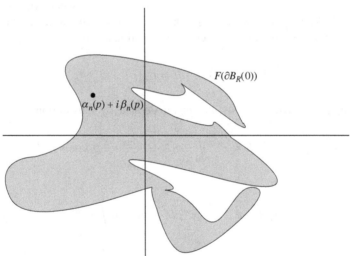

$F(\partial B_R(0))$

$\alpha_n(p) + i\beta_n(p)$

Remark 6.3. An attentive reader may have noticed that the two "Lazer" situations presented in this section can be solved directly by the shooting method. The Landesman–Lazer theorem is deduced as a consequence of what was done in Sect. 1.3.2; Lazer–Leach conditions with $n = 1$ are proposed in Exercise 1.12 (the general case follows in an analogous way). In fact, the same assertion is true for almost all the existence results contained in this book.[1] There are, of course, exceptions: for instance, the shooting method for delay differential equations requires the use of a fixed point in Banach spaces. But, still, we might seriously wonder about

---

[1] In some cases the shooting method seems to be even more effective than degree theory or other methods; see, e.g., [11], where a two-dimensional shooting method is introduced to solve a Neumann problem arising in two-ion electrodiffusion.

the necessity of learning all this stuff if such a simple tool like shooting works so nicely and requires just Bolzano-like theorems.

There are many possible answers to this question. On the one hand, besides the delay equations, in which some idea of shooting can be preserved, the method is no longer applicable when dealing, for example, with partial differential equations, while degree theory retains all its effectiveness. On the other hand, sometimes the degree gives very valuable information about the equation and can be used for other studies such as stability or for proving the multiplicity of solutions, as we shall see in Sect. 6.2.

6.1.4 *An Abstract Continuation Theorem by Mawhin*

A second look at the preceding examples shows that there is a common factor that allows us to unify all the continuation theorems into a single one. This was done in a very general setting by Mawhin, who defined the concept of *coincidence degree*, which is particularly useful for functional equations of the type $Lu = Nu$. We shall introduce only the basic aspects of the theory; for further details see, for example, [41] and [74].

Let X and Y be real normed spaces and $D \subset X$ a subspace, and let $L : D \to Y$ be a linear mapping. L is called a *Fredholm mapping of index* 0 if $Im(L)$ is a closed subspace of Y and

$$dim(Ker(L)) = codim(Im(L)) < \infty.$$

In this case, there exist continuous linear projectors $P : X \to X$ and $Q : Y \to Y$ such that $Im(P) = Ker(L)$ and $Ker(Q) = Im(L)$. Moreover, $Ker(P) \cap Ker(L) = \{0\}$: indeed, if $u \in Ker(L)$, then $u = Pv$ for some v, and hence $Pu = P^2v = Pv = u$. On the other hand, observe that $Lu = L(u - Pu)$ and $u - Pu \in Ker(P)$. This means that the mapping $(L|_{D \cap Ker(P)} : D \cap Ker(P) \to Im(L)$ is bijective, so it has an algebraic inverse K.

Next, consider $N : X \to Y$ continuous, and let $\Omega \subset X$ be bounded and open. We shall say that N is *L-compact* on $\overline{\Omega}$ if the set $QN(\overline{\Omega})$ is bounded and $K(I - Q)N = K \circ (I - Q) \circ N : \overline{\Omega} \to X$ is compact. As in the previous situations, the equation $Lu = Nu$ may be split into the following two equations:

$$u - Pu = KNu, \qquad QNu = 0.$$

Again, the second identity is necessary if we want KNu to be defined, but the difficulty can be avoided if one subtracts QNu to obtain

$$u - Pu = K(I - Q)Nu, \qquad QNu = 0.$$

Remark 6.4. The general procedure of splitting an equation into a system of two equations, the second one finite-dimensional, is known in the literature as a *Lyapunov–Schmidt* reduction.

Finally, observe that if we want to merge the preceding two equations into a single one, some care is needed since QNu and u belong to different spaces. But, very luckily, L is a zero-index Fredholm operator, so there exists an isomorphism $J : Im(Q) = Im(L) \to Ker(L)$, whence the previous system can be equivalently written as

$$u = Pu + JQNu + K(I - Q)Nu.$$

The following continuation theorem is thus obtained (see [74]).

Theorem 6.7. *Let L be a Fredholm mapping of index zero, and let N be L-compact on $\overline{\Omega}$. Assume:*

1. $Lu \neq \lambda Nu$ for $\lambda \in (0,1)$ and $u \in \partial\Omega$.
2. $QNu \neq 0$ for each $u \in Ker(L) \cap \partial\Omega$.
3. $d(JQN, \Omega \cap Ker(L), 0) \neq 0$.

Then the equation $Lu = Nu$ has at least one solution in $D \cap \overline{\Omega}$.

6.2 Ladies and Gentlemen, the Pendulum Is Back

In this section, we shall consider further aspects of the problem studied in Sect. 4.3.2:

$$u''(t) + au'(t) + \sin u(t) = p_0(t) + c, \qquad u(0) = u(1), \, u'(0) = u'(1), \qquad (6.23)$$

with $p_0 : [0,1] \to \mathbb{R}$ continuous such that $\overline{p}_0 = 0$. We have already shown the existence of constants $d \leq D$ such that the problem has at least one solution if and only if $c \in [d, D]$. Following the ideas in [36], we shall prove now by degree theory that if $c \in (d, D)$, then the problem has at least two geometrically different solutions, that is, not differing in a multiple of 2π. To this end, let us firstly prove a preliminary result, which expresses a deep connection between degree theory and the method of upper and lower solutions and is interesting in its own right. We shall state it for the specific problem

$$u''(t) + au'(t) = f(t, u(t)), \qquad u(0) = u(1), \, u'(0) = u'(1), \qquad (6.24)$$

with f continuous, since this is exactly what we need for our application, although the result can be easily extended to different problems and boundary conditions. Suppose that $\alpha < \beta$ are respectively a *strict* lower and an upper solution, that is,

$$\alpha''(t) + a(t)\alpha'(t) > f(t, \alpha(t)), \qquad \beta''(t) + a(t)\beta'(t) < f(t, \beta(t))$$

$$\alpha(0) = \alpha(1), \, \alpha'(0) \geq \alpha'(1) \qquad \beta(0) = \beta(1), \, \beta'(0) \leq \beta'(1),$$

and consider $\Omega := \{u \in C([0,1]) : \alpha(t) < u(t) < \beta(t) \text{ for all } t\}$. Let us recall what we did in Chap. 4 to solve the problem: define the truncation function

$$P(t,u) = \begin{cases} u & \text{if } \alpha(t) \leq u \leq \beta(t), \\ \alpha(t) & \text{if } u < \alpha(t), \\ \beta(t) & \text{if } u > \beta(t), \end{cases}$$

and solve an auxiliary problem

$$u''(t) + au'(t) - u(t) = f(t, P(t, u(t))) - P(t, u(t)),$$

$$u(0) = u(1), u'(0) = u'(1).$$

Using Schauder's theorem, it is seen that, this problem has at least one solution, which lies between α and β, and hence solves the original problem. Now observe that the truncation function does not affect the elements of $\overline{\Omega}$, so what we have proven, in fact, is that the compact operator $K : C([0,1]) \to C([0,1])$ given by $Kv = u$, the unique solution of

$$u''(t) + au'(t) - u(t) = f(t, v(t)) - v(t), \qquad u(0) = u(1), u'(0) = u'(1), \quad (6.25)$$

has a fixed point in $\overline{\Omega}$. So why use the truncation function?

The answer is very simple: if we want to apply Schauder's theorem, then we need to find a closed convex and bounded subset of $C([0,1])$ that is left invariant under K, and this cannot be a priori guaranteed for arbitrary f. Truncation converts the right-hand side of the equation into a bounded function; hence, Schauder's theorem applies for a sufficiently large ball, and the couple (α, β) does the rest of the work.

The following result reveals the strength of degree theory. It ensures the existence of fixed points even in some situations in which we might fail in advance to prove the existence of invariant regions.

Theorem 6.8. *Let $\alpha < \beta$ be respectively a strict lower solution and a strict upper solution of (6.24). Then $deg(I - K, \Omega, 0) = 1$.*

Proof. Consider the previous truncation function P, and let $K_P : C([0,1]) \to C([0,1])$ be defined by $K_P v = u$, the unique solution of (6.25). Let u be a fixed point of K_P, and suppose, for example, that $u(t_0) \geq \beta(t_0)$ for some $t_0 \in [0,1)$. We may assume that $u - \beta$ achieves its absolute maximum at t_0; hence, $u'(t_0) = \beta'(t_0)$ and $u''(t_0) \leq \beta''(t_0)$. Moreover, $P(t_0, u(t_0)) = \beta(t_0)$, so

$$u''(t_0) + a(t_0)u'(t_0) = f(t_0, \beta(t_0)) > \beta''(t_0) + a(t_0)\beta'(t_0),$$

a contradiction. Thus, $u(t) < \beta(t)$ for all t, and in the same way it is proven that $\alpha < u$. In particular, K_P does not vanish outside Ω; thus, $deg(I - K_P, \Omega, 0)$ is defined, and clearly it coincides with $deg(I - K, \Omega, 0)$. Moreover, $Im(K_P)$ is bounded; hence, for large enough R the homotopy $I - \lambda K_P$ does not vanish on $\partial B_R(0)$ for $\lambda \in [0,1]$. From the homotopy invariance and excision property,

$$deg(I - K_P, \Omega, 0) = deg(I - K_P, B_R(0), 0) = deg(I, B_R(0), 0) = 1,$$

and the proof is complete. □

As a direct application, let us prove the existence of a second solution of (6.23) when c is an interior point of $[d,D]$.[2] Specifically, suppose that u_1 and u_2 are solutions for some $c_1 < c_2$. By the periodicity of the sine function, we may assume that

$$u_2 \leq u_1, \qquad u_2 + 2\pi \nleq u_1,$$

that is, $u_2(t) \leq u_1(t)$ for all t, and there exists t_0 such that $u_1(t_0) < u_2(t_0) + 2\pi$. Then u_2 and u_1 are respectively strict lower and upper solutions of (6.23). We claim that $u_2(t) < u_1(t)$ for all t; indeed, otherwise the function $u_1 - u_2$ would achieve its absolute minimum at some $t_0 \in [0,1)$ with $u_1(t_0) = u_2(t_0)$, $u_1'(t_0) = u_2'(t_0)$ and $u_1''(t_0) \geq u_2''(t_0)$, so

$$u_1''(t_0) + a(t_0)u_1'(t_0) + \sin(u_1(t_0)) \geq u_2''(t_0) + a(t_0)u_2'(t_0) + \sin(u_2(t_0)).$$

In other words, $p(t_0) + c_1 \geq p(t_0) + c_2$, a contradiction. Now take $\alpha = u_2$ and $\beta = u_1$; then from the previous theorem we know that $deg(I - K, \Omega, 0) = 1$. But that's not all! Observe, in the first place, that the same result is valid if we instead consider $\alpha = u_2 + 2\pi$ and $\beta = u_1 + 2\pi$: this does not constitute any novelty because of the periodicity. But, further, we may also consider $\alpha = u_2$ and $\beta = u_1 + 2\pi$. To keep things more or less organized, let us define the sets

$$\Omega_1 := \{u \in C([0,1]) : u_2 < u < u_1\},$$

$$\Omega_2 := \{u \in C([0,1]) : u_2 + 2\pi < u < u_1 + 2\pi\},$$

and

$$\Omega := \{u \in C([0,1]) : u_2 < u < u_1 + 2\pi\}.$$

Clearly, Ω_1 and Ω_2 are disjoint subsets of Ω, so from the additivity property we may decompose $deg(I - K, \Omega, 0)$ as

$$deg(I - K, \Omega_1, 0) + deg(I - K, \Omega_2, 0) + deg(I - K, \Omega \backslash (\overline{\Omega_1 \cup \Omega_2}), 0).$$

The first two terms are equal to 1, as well as $deg(I - K, \Omega, 0)$, so we deduce that

$$deg(I - K, \Omega \backslash (\overline{\Omega_1 \cup \Omega_2}), 0) = -1.$$

In particular, K has at least one fixed point in $\Omega \backslash (\overline{\Omega_1 \cup \Omega_2})$.

It is worth recalling, from the previous problem, that the role of periodicity was essential in order to produce a well-ordered couple of a lower and an upper solution: for given $c_1 < c_2$ belonging to $I(p_0)$ we just picked up arbitrary solutions u_1 and u_2 and then took advantage of the fact that $u_1 + 2k\pi$ is also a solution for c_1, for all $k \in \mathbb{Z}$. But one has the right to wonder if this procedure is still possible if we replace the sine function by an arbitrary bounded function g, namely, the problem

---

[2] In the delightful paper [95], an upper bound for the number of solutions is obtained by means of a classical result in complex analysis: the Jensen inequality.

$$u''(t) + au'(t) + g(u(t)) = p_0(t) + c, \qquad u(0) = u(1), u'(0) = u'(1), \qquad (6.26)$$

with $\overline{p}_0 = 0$. Straightforward imitation of what we did in the first part of Sect. 4.3.2 proves the existence of a bounded nonempty set $I(p_0) \subset \mathbb{R}$ such that (6.26) has solutions if and only if $c \in I(p_0)$. But what does $I(p_0)$ look like? Note that $I(p_0)$ is not necessarily closed; for example, according to the Landesman–Lazer theorem, the equation

$$u''(t) + \arctan u(t) = p_0(t) + c$$

has periodic solutions if and only if $c \in \left(-\frac{\pi}{2}, \frac{\pi}{2}\right)$.

The novelty we would like to point out now is that, regardless of whether or not g is periodic, $I(p_0)$ is always an interval. And the reason is very simple: if $c_1 < c_2$ belong to $I(p_0)$ and $c \in (c_1, c_2)$, then take the respective solutions u_1 and u_2, so (u_2, u_1) is a couple of a lower and an upper solution for c.

Hmmm... wait one moment! Did anyone hear *well-ordered* couple? The whole story looks quite suspicious, so before the reader starts complaining, we must make a very firm declaration: we don't care whether or not $u_2 \leq u_1$. This is a consequence of an existence result for the general case of a bounded nonlinearity that, adapted to this case, reads as follows.

Theorem 6.9. *Let* $f : [0,1] \times \mathbb{R} \to \mathbb{R}$ *be continuous and bounded, and assume that* $\alpha, \beta \in C^2([0,1])$ *satisfy*

$$\alpha''(t) + a\alpha'(t) \geq f(t, \alpha(t)), \qquad \beta''(t) + a\beta'(t) \leq f(t, \beta(t)),$$

$$\alpha(0) - \alpha(1) = \beta(0) - \beta(1) = 0, \qquad \alpha'(0) - \alpha'(1) \geq 0 \geq \beta'(0) - \beta'(1).$$

Then the problem

$$u''(t) + au'(t) = f(t, u(t)), \qquad u(0) = u(1), u'(0) = u'(1) \qquad (6.27)$$

has at least one solution.

Note that no order relation between α and β is assumed, and, obviously, the result provides no information about the location of such a solution. The argument is somewhat subtle: if one supposes the problem has no solutions, then it is shown that an ordered couple of a lower and an upper solution exists, thereby producing a contradiction. We sketch a proof here; details are left to the reader. Let

$$X := \{u \in C([0,1]) : u(0) = u(1)\}, \qquad \tilde{X} := \{u \in X : \overline{u} = 0\},$$

and let $K : \tilde{X} \to X$ be the compact operator given by $K\varphi = u$, where u is the unique solution of the problem

$$u''(t) + au'(t) = \varphi(t), \qquad u(0) = u(1), \quad u'(0) = u'(1), \quad \overline{u} = 0.$$

For $r \in \mathbb{R}$, consider the integro-differential problem

$$u''(t) + au'(t) = f(t,u(t)) - \int_0^1 f(s,u(s))\,ds, \qquad u(0) = u(1), \ \overline{u} = r. \qquad (6.28)$$

Using Schauder's theorem, for each $r \in \mathbb{R}$ there exists at least one solution u, which also satisfies $u'(0) = u'(1)$. We look, for some r, for a solution of (6.28) such that $\int_0^1 f(s,u(s))\,ds = 0$. Define

$$Nu := f(\cdot,u), \qquad \tilde{N}u := Nu - \overline{Nu};$$

then $u \in X$ is a solution of (6.27) if and only if $u - \overline{u} = K\tilde{N}u$ and $\overline{Nu} = 0$. It proves convenient to identify X with $\mathbb{R} \times \tilde{X}$ by writing each $u \in X$ as a pair $u = (\overline{u}, \tilde{u})$, where $\tilde{u} = u - \overline{u}$. Let $\mathscr{A}$ be the set of those $(r,\tilde{u})$ such that $r + \tilde{u}$ solves (6.28), that is,

$$\mathscr{A} := \{(r,\tilde{u}) \in \mathbb{R} \times \tilde{X} : \tilde{u} = K\tilde{N}(r+\tilde{u})\};$$

then the boundedness of f implies that $\mathscr{A} \subset \mathbb{R} \times B_R(0)$ for some large enough R. Observe, furthermore, that if u solves (6.28), then $\|\tilde{u}'' + a\tilde{u}'\|_\infty \le 2\|f\|_\infty$, and hence, by the Arzelá–Ascoli theorem, the projection of $\mathscr{A}$ into $\tilde{X}$ is compact. From the previous considerations, given $s \in \mathbb{R}$, the set

$$\mathscr{A}^s := \{(r,\tilde{u}) \in \mathscr{A} : r = s\}$$

is nonempty and compact. We claim that for any $a < b$ there exists a connected set $C \subset \mathscr{A}$ that intersects both $\mathscr{A}^a$ and $\mathscr{A}^b$. This requires an important result from the theory of abstract metric spaces, known as the *Whyburn lemma* (see [108]).

Lemma 6.1. *(Whyburn) Let X be a metric space, and let $K \subset X$ be compact. If K_0 and K_1 are disjoint closed subsets of K such that no connected subset of K intersects both of them, then there exist disjoint closed $C_0, C_1 \subset K$ such that $K_0 \subset C_0$, $K_1 \subset C_1$ and $K = C_0 \cup C_1$. Furthermore, there exists an open bounded set Ω containing C_0 such that $\Omega \cap C_1 = \emptyset$ and $\partial\Omega \cap K = \emptyset$.*

To prove the claim, let us suppose that $\mathscr{A}^a$ and $\mathscr{A}^b$ are disconnected and define the (also compact) set $\mathscr{K} := \bigcup_{a \le s \le b} \mathscr{A}^s = \{(r,\tilde{u}) \in \mathscr{A} : a \le r \le b\}$. Now take $C_a \supset \mathscr{A}^a$, $C_b \supset \mathscr{A}^b$ closed disjoint sets such that $\mathscr{K} = C_a \cup C_b$ and $\Omega \subset \mathbb{R} \times B_R(0)$ containing C_a such that $\Omega \cap C_b = \partial\Omega \cap \mathscr{K} = \emptyset$. Let $\Omega_s := \{\tilde{u} : s + \tilde{u} \in \Omega\} \subset B_R(0)$ and consider the mapping $F : \overline{\Omega} \to \tilde{X}$ given by $F(s,\tilde{u}) := \tilde{u} - K\tilde{N}(s+\tilde{u})$. If $F(s,\tilde{u}) = 0$ for some $(s,\tilde{u}) \in \partial\Omega$, then $(s,\tilde{u}) \notin \mathscr{K}$, which contradicts the definition of $\mathscr{K}$. By the homotopy invariance mentioned in Remark 5.4, $deg(F(s,\cdot), \Omega_s, 0)$ does not depend on s. Moreover, observe the following properties:

1. $\tilde{u} \ne \sigma K\tilde{N}(a+\tilde{u})$ for $\tilde{u} \in \partial B_R(0)$ and $\sigma \in [0,1]$ since $Im(K\tilde{N}) \subset B_R(0)$.
2. $F(a,\cdot)$ does not vanish outside Ω_a since $\mathscr{A}^a \subset \Omega_a$.
3. $F(b,\cdot)$ does not vanish in Ω_b since $\Omega \cap C_b = \emptyset$.

The preceding properties imply the following sequence of equalities:

$$1 = deg(F(a,\cdot), B_R(0), 0) = deg(F(a,\cdot), \Omega_a, 0) = deg(F(b,\cdot), \Omega_b, 0) = 0.$$

This contradiction proves the existence of a connected subset $C \subset \mathcal{K}$ such that $C \cap \mathcal{A}^a \neq \emptyset$ and $C \cap \mathcal{A}^b \neq \emptyset$.

At this point, the reader might appreciate it if we gave at least an idea of what we are doing. To this end, recall that u is a solution of the problem if and only if $(\bar{u}, \tilde{u}) \in \mathcal{A}$ and $\overline{N(\bar{u} + \tilde{u})} = 0$. So our plan consists in showing that the function $(r, \tilde{u}) \mapsto \overline{N(r + \tilde{u})}$ vanishes in $\mathcal{A}$.

In the first place, observe that if $(r, \tilde{u}) \in \mathcal{A}$ and $\overline{N(r + \tilde{u})} > 0$, then $u = r + \tilde{u}$ satisfies

$$f(\cdot, u) = N(r + \tilde{u}) > N(r + \tilde{u}) - \overline{N(r + \tilde{u})} = \tilde{N}(r + \tilde{u}).$$

Because $(r, \tilde{u}) \in \mathcal{A}$, we know that $K \tilde{N}(r + \tilde{u}) = \tilde{u}$ or, equivalently, $\tilde{u}'' = \tilde{N}(r + \tilde{u})$. Hence $f(\cdot, u) > \tilde{N}(r + \tilde{u}) = \tilde{u}'' = u''$, so u is an upper solution. In the same way, if $\overline{N(r + \tilde{u})} < 0$, then u is an upper solution.

Now suppose that $\overline{N(r + \tilde{u})} \neq 0$ for all $(r, \tilde{u}) \in \mathcal{A}$, and take $a < b$ such that

$$b > R + \max_{t \in [0,1]} \alpha(t), \qquad a < -R + \min_{t \in [0,1]} \beta(t).$$

Let $C \subset \mathcal{A}$ be, as before, a connected set that meets both $\mathcal{A}^a$ and $\mathcal{A}^b$. By continuity, the sign of $\overline{N(r + \tilde{u})}$ is the same for all $(r, \tilde{u}) \in C$. If it is positive, then take $\tilde{u}$ such that $(b, \tilde{u}) \in C$; thus, $u(t) := b + \tilde{u}(t)$ is an upper solution with $u(t) > b - R > \alpha(t)$ for all t, and we deduce the existence of a solution between α and u. An analogous conclusion is obtained if $\overline{N(r + \tilde{u})} < 0$ for all $(r, \tilde{u}) \in C$.

6.3 Not So Far, But Still Good

In Sect. 6.1.3 we described some of the most famous Landesman–Lazer-type results, with the warning that they would be improved. In some sense, it is rather evident that the original conditions are not the best one can get: for instance, a rapid checklist of what is needed for a proof shows that conditions (6.18) and (6.19) may be replaced without any damage by

$$\limsup_{u \to -\infty} g(u) < \bar{p} < \liminf_{u \to +\infty} g(u)$$

and

$$\limsup_{u \to +\infty} g(u) < \bar{p} < \liminf_{u \to -\infty} g(u),$$

respectively. The explanation is very simple: when, as before, one supposes that u_n is a solution for some $\lambda_n \in (0, 1)$ with $\|u_n\|_\infty \to \infty$, then $\|u_n - \overline{u_n}\|_\infty$ is bounded and

$$\int_0^1 g(u_n(t)) \, dt = \bar{p}. \tag{6.29}$$

Now it is true that the left-hand side may not converge, but one can still use Fatou's lemma to obtain inequalities that contradict (6.29). However, even more general cases are easily obtained; for example, we mentioned a result by Lazer [68] in which the existence of solutions is proven under the weaker hypotheses

$$g(-u) < \overline{p} < g(u) \qquad \text{for } u \geq R_0 \tag{6.30}$$

or

$$g(u) < \overline{p} < g(-u) \qquad \text{for } u \geq R_0. \tag{6.31}$$

In particular, the so-called *vanishing nonlinearities* are not necessarily forbidden: we need only that the function $g(u) - \overline{p}$ has different signs when $u \gg 0$ and $u \ll 0$, so it might perfectly happen that $\lim_{|u| \to \infty} g(u) - \overline{p} = 0$. We recall that, in the original result by Lazer, the previous inequalities are nonstrict; this extension can be deduced from the strict case by an approximation argument.

However, we shall see that all the previous results admit further extensions; in fact, it is possible to establish Landesman–Lazer-type conditions in which no specific asymptotic behavior for g is assumed. With this aim, let us review the main steps in our proof: firstly, we have obtained bounds for the problem

$$u''(t) = \lambda (p(t) - g(u(t))), \qquad u(0) = u(1), \, u'(0) = u'(1),$$

with $\lambda \in (0,1)$. More specifically, we have shown that

$$\|u - \overline{u}\|_\infty < \frac{\|p\|_\infty + \|g\|_\infty}{4\pi} := r$$

and, immediately after, concluded that $\overline{u}$ is also bounded from the identity

$$\int_0^1 g(u(t)) \, dt = \overline{p}.$$

A deeper analysis reveals that the last step does not require taking limits: because $u(t)$ belongs to the interval $(\overline{u} - r, \overline{u} + r)$ for all t, the preceding equality forces g to take the value $\overline{p}$ at least once over this interval. Thus, we would already be very happy with a condition like

$$g(u) \neq \overline{p} \qquad \text{if } u \in (-R - r, -R + r) \cup (R - r, R + r)$$

for some R. But suddenly we recall that there is also a degree condition to be fulfilled, so we need to reformulate the assumption to guarantee a change of sign, namely,

$$g(-u) < \overline{p} < g(u) \qquad \text{if } u \in (R - r, R + r),$$

or vice versa. Finally, we may also wise up to the fact that the role of $\pm R$ can be played by arbitrary real numbers a, b, with the only obvious condition that the distance between them must be at least $2r$. Thus, a general nonasymptotic Landesman–Lazer condition reads as follows:

$$g(u) < \overline{p} \text{ if } u \in (a - r, a + r) \qquad \text{and} \qquad g(u) > \overline{p} \text{ if } u \in (b - r, b + r). \tag{6.32}$$

Note that both possible Landesman–Lazer conditions are expressed at once since we did not prescribe any ordering between a and b. Under this condition, the continuation theorem can be applied as before with a slightly different choice of the domain:

$$\Omega = \{u \in C([0,1]) : \|u - \bar{u}\|_\infty < r, \bar{u} \in D\}, \tag{6.33}$$

where $D = (a,b)$ or $D = (b,a)$ according to the two cases $a < b$ or $b < a$.

But the result says, in fact, much more; indeed, any time we have intervals I, J of length $2r$ such that $g - \bar{p}$ has different signs over I and J, there exists a solution u between I and J, in the sense that $u(t)$ lies the whole time in the smallest interval containing $I \cup J$. In particular, g may oscillate around $\bar{p}$ along the whole line, a situation that cannot occur in the Landesman–Lazer theorem or in the more general result in [68]: if the positivity and negativity intervals are long enough (namely, of length at least equal to $2r$), then the existence of infinitely many solutions is deduced.

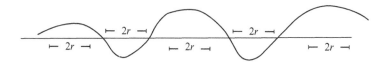

This makes us think over some nonexistence results, for example those obtained in [94] and more generally in [97] for the pendulum equation

$$u''(t) + au'(t) + \sin u(t) = p(t).$$

Under some choices of a and the T-periodic function p with zero average, the problem has no T-periodic solutions, although the nonlinearity $g(u) = \sin u$ has infinitely many intervals of positivity and negativity of length π. Why? The reason is very simple: the corresponding r is larger than $\frac{\pi}{2}$, so the previous result does not apply. In other words, the nonlinearity oscillates too fast. The situation is different in the case of a so-called *expansive nonlinearity* [38]: for example, the equation

$$u''(t) + \sin \sqrt[3]{u(t)} = p(t)$$

has infinitely many T periodic solutions for any T-periodic p such that $|\bar{p}| < 1$. Indeed, here the set of positivity of the function $\sin \sqrt[3]{u} - \bar{p}$ is a union of intervals

$$\bigcup_{k \in \mathbb{Z}} \left((\alpha + 2k\pi)^3, (\pi - \alpha + 2k\pi)^3 \right)$$

for some $\alpha \in (-\pi, \pi)$, and no matter how large $r = \frac{\|p\|_\infty + 1}{4\pi}$ is, the length of the intervals exceeds $2r$ when $|k|$ is large enough. The same conclusion is obviously true for the negativity set, so the result follows.

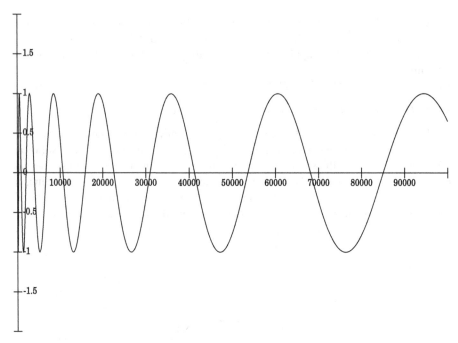

Regarding the Lazer–Leach conditions, the situation turns more delicate. Again, by Fatou's lemma it is readily seen that Theorem 6.6 is still true if the quantity $|g(+\infty) - g(-\infty)|$ is replaced by

$$\liminf_{u \to +\infty} g(u) - \limsup_{u \to -\infty} g(u)$$

or

$$\liminf_{u \to -\infty} g(u) - \limsup_{u \to +\infty} g(u),$$

but in view of the previous case, we want more.

For instance, the existence of limits is not necessary, as was proven by Fabry and Franchetti in [34], where conditions analogous to (6.30) and (6.31) are obtained:

$$\frac{4}{\pi} g(u) > \sqrt{\alpha_m(p)^2 + \beta_m(p)^2} \quad \text{and} \quad \frac{4}{\pi} g(-u) < -\sqrt{\alpha_m(p)^2 + \beta_m(p)^2} \quad (6.34)$$

(or vice versa) for $u \geq R_0$. It is a simple exercise to verify that these conditions represent, indeed, a generalization of the Lazer–Leach result: it suffices to observe that a constant may be added to both sides of the equation, and the m-th Fourier coefficients of the right-hand-side term remain the same. Thus, if, for example, the first Lazer–Leach condition holds, that is,

$$\frac{\pi}{2} \sqrt{\alpha_m(p)^2 + \beta_m(p)^2} < \liminf_{u \to +\infty} g(u) - \limsup_{u \to -\infty} g(u),$$

then it may be assumed without loss of generality that

$$\liminf_{u \to +\infty} g(u) = -\limsup_{u \to -\infty} g(u),$$

and the Fabry–Franchetti condition is also satisfied.

Remark 6.5. At first sight, one might expect that a natural generalization of (6.30) or (6.31) to the present case could be

$$\frac{2}{\pi}[g(u) - g(v)] > \sqrt{\alpha_m(p)^2 + \beta_m(p)^2} \tag{6.35}$$

for all $u \geq R_0$ and $v \leq -R_0$ (or vice versa). But if we assume as before that $\liminf_{u \to +\infty} g(u) = -\limsup_{u \to -\infty} g(u)$, then condition (6.35) implies that

$$\frac{4}{\pi}g(u) \geq \sqrt{\alpha_m(p)^2 + \beta_m(p)^2} \text{ and } \frac{4}{\pi}g(v) < -\sqrt{\alpha_m(p)^2 + \beta_m(p)^2}$$

or

$$\frac{4}{\pi}g(u) > \sqrt{\alpha_m(p)^2 + \beta_m(p)^2} \text{ and } \frac{4}{\pi}g(v) \leq -\sqrt{\alpha_m(p)^2 + \beta_m(p)^2}$$

for all $u \geq R_0$ and $v \leq -R_0$. Nonetheless, these conditions do not necessarily imply the existence of solutions since it is not known if the result in [34] is still valid when the inequalities in (6.34) are nonstrict. The approximation argument used for the nonstrict case of (6.30), which consists in adding, for example, a term $\frac{1}{n} \arctan u(t)$ to the equation, fails for this case since the bounds for the corresponding solution u_n depend on each n.

It is worth noting that, although the conditions on the limits of g were removed in [34], some behavior of g at infinity is still required. In some sense, this could be expected since the assumptions are stated only in terms of the projection of p: if g satisfies (6.34), then the existence of solutions still holds when p is replaced by any other function with the same mth Fourier coefficients. In [8], it was shown that if some other data of p is prescribed (more specifically, its norm), then it is possible to give a nonasymptotic condition that allows arbitrary behavior of g outside a compact set. More precisely, for $R, \varepsilon > 0$, a nontrivial interval $I = I(R, \varepsilon) \subset (0, +\infty)$ and a function $C = C(R, \varepsilon)$ are defined such that if

$$\frac{2\sqrt{1 - \varepsilon^2}}{\pi} \left(\inf_{u \in I} g(u) - \sup_{-v \in I} g(v) \right) - C > \sqrt{\alpha_m(p)^2 + \beta_m(p)^2} \tag{6.36}$$

or

$$\frac{2\sqrt{1 - \varepsilon^2}}{\pi} \left(\inf_{-v \in I} g(v) - \sup_{u \in I} g(u) \right) - C > \sqrt{\alpha_m(p)^2 + \beta_m(p)^2}, \tag{6.37}$$

then the problem admits at least one solution. Moreover,

$$C(R, \varepsilon) \to 0 \qquad \text{as } \varepsilon \to 0 \text{ and } R \to +\infty,$$

so the original Lazer-Leach conditions are easily retrieved from (6.36) and (6.37). The interval I depends not only on R and ε but also on $\|g\|_\infty$ and $\|p\|_{L^2}$, as well as the function C. Observe that (6.36) and (6.37) establish some behavior of g over the set $I \cup -I$, which is symmetric with respect to the origin; in this sense, the conditions are more restrictive than (6.32), where we did not need to assume $a = -b$. Nonetheless, the conditions are sufficiently good for obtaining some multiplicity results in this case, too; for example, if (6.36) and (6.37) hold for some different choices of R and ε, then the degree of the associated operator shall be 1 in one case and -1 in the other one, so by excision one obtains at least two different solutions.

6.3.1 Attention Please: g May Be Unbounded

Our previous considerations on the Landesman–Lazer result and its generalizations were based on the fact that g is bounded; specifically, the value of r depends on $\|g\|_\infty$ and $\|p\|_\infty$. However, in some cases the boundedness assumption may be relaxed, or even dropped. For instance, this is the case when the equation contains a nonzero term $au'(t)$: as was already stated once upon a time in Chap. 1, accurate bounds for u' are obtained directly from the problem

$$u''(t) + au'(t) = \lambda\,(p(t) - g(u(t))), \qquad u(0) = u(1),\ u'(0) = u'(1).$$

The forgetful reader may appreciate it if we evoked the procedure once again. Just multiply by $u'(t)$ and integrate to deduce

$$\|u'\|_{L^2}^2 = \frac{\lambda}{a} \int_0^1 p(t)u'(t)\,dt \le \frac{1}{|a|}\|p\|_{L^2}\|u'\|_{L^2}.$$

Thus, $\|u'\|_{L^2} \le \frac{1}{|a|}\|p\|_{L^2}$, and the proof follows since $\|u - \overline{u}\|_\infty \le \frac{1}{2}\|u'\|_{L^2}$.

Also, as "proven" in Exercise 5.21, the boundedness condition is not required in the easy case of Landesman–Lazer, that is, $g(-u) > \overline{p} > g(u)$ for $u \gg 0$. This can be expressed in a more meaningful way:

$$(g(u) - \overline{p})u < 0 \qquad \text{for } |u| \text{ large enough.}$$

However, this condition already implies the nonasymptotic condition (6.32) for some $a \gg 0 \gg b$. Nevertheless, the weaker condition

$$(g(u) - \overline{p})u \le K \qquad \text{for all } u$$

for some $K \ge 0$ is still enough to obtain bounds for $\|u - \overline{u}\|_\infty$. Indeed, if u satisfies the periodic conditions and

$$u''(t) = \lambda\,(p(t) - g(u(t)))$$

for some $\lambda \in (0, 1)$, then

$$-\int_0^1 u''(t)u(t)\,dt = \lambda \int_0^1 (\overline{p} - p(t) + g(u(t)) - \overline{p})\,dt < \lambda \int_0^1 (\overline{p} - p(t))u(t)\,dt + K.$$

Since $\int_0^1 (\overline{p} - p(t))u(t)\,dt = \int_0^1 (\overline{p} - p(t))(u(t) - \overline{u})\,dt$, integration by parts and Cauchy–Schwarz and Wirtinger inequalities yield

$$\int_0^1 u'(t)^2\,dt < \|p - \overline{p}\|_{L^2}\|u - \overline{u}\|_{L^2} + K \le \frac{1}{2\pi}\|p - \overline{p}\|_{L^2}\|u'\|_{L^2} + K.$$

This provides a bound for $\|u'\|_{L^2}$ and, consequently, for $\|u - \overline{u}\|_\infty$. The preceding condition may be replaced by an even weaker one:

$$(g(u) - \overline{p})u \le c|u|^\gamma + K \qquad \text{for all } u \tag{6.38}$$

for some $K, c > 0$, and some $\gamma < 2$, although in this case the values a, b in (6.32) and r are not independent (Exercise 6.7). The same thing happens with the sublinearity condition or, more generally,

$$|g(u)| \le c|u| + K \qquad \text{for all } u.$$

Here the equality $u'' = \lambda(p - g(u))$ with $0 < \lambda < 1$ implies

$$4\pi\|u - \overline{u}\|_\infty \le \|u''\|_\infty < \|p\|_\infty + c\|u\|_\infty + K \le \|p\|_\infty + c\|u - \overline{u}\|_\infty + K + c|\overline{u}|,$$

and thus

$$(4\pi - c)\|u - \overline{u}\|_\infty < \|p\|_\infty + K + c|\overline{u}|.$$

Assume that $c < 4\pi$; then

$$\|u - \overline{u}\|_\infty \le M + \frac{c}{4\pi - c}|\overline{u}|,$$

where $M = \frac{\|p\|_\infty + K}{4\pi - c}$. In particular, if $c < 2\pi$, then $\frac{c}{4\pi - c} < 1$; thus, if R is large enough, then $r := M + \frac{cR}{4\pi - c} < R$, and it suffices to assume (6.32) with $a = -R$ and $b = R$, or vice versa. Obviously, conditions with more general values of a and b are also possible.

We end this section with another typical situation: g is bounded, but only from one side. For example, assume that $g(u) \ge -K$ for all u and some constant K. Now the previous equality can be written as

$$u''(t) = \lambda(p(t) - g(u(t))) = \lambda(p(t) + K) - \lambda(g(u(t)) + K),$$

and hence

$$|u''(t)| < |p(t) + K| + g(u(t)) + K.$$

Because $\int_0^1 g(u(t))\,dt = \overline{p}$, we deduce

$$\int_0^1 |u''(t)|\,dt < \int_0^1 |p(t) + K|\,dt + \overline{p} + K,$$

so

$$\|u - \overline{u}\|_\infty \le \frac{1}{2}\|u'\|_\infty \le \frac{1}{4}\int_0^1 |u''(t)|\,dt < \frac{1}{4}\left(\int_0^1 |p(t) + K|\,dt + \overline{p} + K\right) := r.$$

As the reader might imagine, here the Lazer–Leach situation is again more subtle. Surprisingly, it seems that sublinearity no longer suffices, and a "natural" growth condition for g is the following one, already obtained in [34]:

$$\lim_{|u| \to \infty} \frac{g(u)}{|u|^\alpha} = 0 \tag{6.39}$$

for some $\alpha \in [0, \frac{1}{2})$. The sharpness of this condition is somehow explained in [8], where it is shown that $C(R, \varepsilon) = O(R^{2\alpha - 1})$ for $\varepsilon \ll R^{-\alpha}$.

Another remarkable difference with the Landesman-Lazer case is that, as mentioned, no growth conditions at all are necessary under condition (6.31). This can also be seen from the variational formulation of the problem. However, none of these arguments is valid anymore when Leach is invited to join the party.

6.4 When the Geometry Goes Marching In

As we have already seen (and repeated many times!), Landesman–Lazer conditions are very important in resonant problems, as long as one must deal with a one-dimensional kernel. The Lazer–Leach situation was more complicated since the projection to the kernel involves the study of oscillatory integrals. But for the moment, we did not talk about *systems* of equations. So let us complicate our lives a little bit.

We may believe that problem (6.17) with $g \in C(\mathbb{R}^n, \mathbb{R}^n)$ continuous and bounded and $p \in C([0,1])$ is still simple, in the sense that resonance is produced at the first eigenvalue $\lambda_0 = 0$. However, now the kernel of the associated linear operator $Lu = u''$, namely the set of constants, is n-dimensional. Is there any "natural" extension of Landesman–Lazer conditions in this case?

Recall our most elementary interpretation of the result for $n = 1$: if $\overline{p}$ lies between the limits $g(\pm\infty)$, then the problem has at least one solution. This may be regarded as a condition over the "sphere" $S^0 = \{-1, 1\}$: if for $v = \pm 1$ we set $g_{\pm 1} = g(\pm\infty)$, then the function $\theta : S^0 \to S^0$ given by $\theta(v) = \frac{g_v - \overline{p}}{|g_v - \overline{p}|}$ is well defined, and furthermore, any continuous extension $\tilde{\theta} : [-1, 1] \to \mathbb{R}$ of θ has a nonzero Brouwer degree. But we may express this fact simply by saying that $deg(\theta) \ne 0$, where the degree is understood in the following sense.

Definition 6.1. Let S^{n-1} be the unit sphere of $\mathbb{R}^n$, let $\theta : S^{n-1} \to S^{n-1}$ be continuous, and let $\tilde{\theta} : \overline{B_1(0)} \to \mathbb{R}^n$ be a continuous extension of θ. Then the degree of θ is defined as

$$deg(\theta) := deg(\tilde{\theta}, B_1(0), 0).$$

It is clear that $deg(\tilde{\theta}, B_1(0), 0)$ does not depend on the choice of $\tilde{\theta}$, so $deg(\theta)$ is well defined. This definition is equivalent to the standard one, which can be easily introduced using homology theory.

From the previous considerations, the following result, adapted from a theorem proven by Nirenberg in [78], may be regarded as a natural extension of the Landesman–Lazer theorem.

Theorem 6.10. *Let $p \in C([0,1])$, let $g \in C(\mathbb{R}^n, \mathbb{R}^n)$ be bounded, and assume that the radial limits*

$$g_v := \lim_{s \to +\infty} g(sv)$$

exist uniformly with respect to $v \in S^{n-1}$. Then problem (6.17) has at least one solution if the following conditions hold:

- $g_v \neq \overline{p}$ *for any $v \in S^{n-1}$;*
- $deg(\theta) \neq 0$, *where $\theta : S^{n-1} \to S^{n-1}$ is the mapping defined by $\theta(v) = \frac{g_v - \overline{p}}{|g_v - \overline{p}|}$.*

Then the problem

$$u''(t) + g(u(t)) = p(t), \qquad u(0) = u(1), \quad u'(0) = u'(1) \qquad (6.40)$$

has at least one solution.

After all the work done for the previous cases, the proof of this result is really straightforward. It is worth verifying, in the first place, that θ is a continuous mapping, and thus its degree is defined: indeed, for $\varepsilon > 0$ fix s such that $|g_v - g(sv)| < \frac{\varepsilon}{3}$ for all $v \in S^{n-1}$. For arbitrary $w \in S^{n-1}$, choose $\delta > 0$ such that $|g(sv) - g(sw)| < \frac{\varepsilon}{3}$ for $|v - w| < \delta$; then

$$|g_v - g_w| < |g_v - g(sv)| + |g(sv) - g(sw)| + |g(sw) - g_w| < \varepsilon,$$

and the proof follows.

Next, we may proceed with the Nirenberg result. As before, if

$$u_k''(t) = \lambda_k(p(t) - g(u_k(t))), \qquad u_k(0) = u_k(1), \ u_k'(0) = u_k'(1)$$

for some $\lambda_k \in (0,1)$ and $\|u_k\|_\infty \to \infty$, then $\|u_k - \overline{u}_k\|_\infty < \frac{\|p\|_\infty + \|g\|_\infty}{4\pi}$ and $|\overline{u}_k| \to \infty$. Taking a subsequence if necessary, it may be assumed that $\frac{\overline{u}_k}{|\overline{u}_k|}$ converges to some $v \in S^{n-1}$, so it is readily seen that the functions $\frac{u_k}{|u_k|}$ converge uniformly to v. Hence

$$\overline{p} = \int_0^1 g(u_k(t)) \, dt = \int_0^1 g\left(|\overline{u}_k| \frac{u_k(t)}{|u_k|}\right) dt \to g_v,$$

a contradiction.

Finally, the second Nirenberg condition implies $deg(g - \overline{p}, B_R(0), 0) \neq 0$ for R large enough; for example, observe that $\overline{p} - g$ does not vanish outside a ball $B_{R_0}(0)$, so, by excision, $deg(\overline{p} - g, B_R(0), 0)$ is well defined and independent of R when $R \geq R_0$. Now let $\theta_R(u) : S^{n-1} \to S^{n-1}$ be given by $\theta_R(u) := \frac{g(Ru) - \overline{p}}{|g(Ru) - \overline{p}|}$; then $\theta_R \to \theta$ uniformly, so $deg(\theta_R) \to deg(\theta)$. Moreover, the homotopy $\lambda \theta_R(u) + (1 - \lambda)(g(Ru) - \overline{p})$ does not vanish for $\lambda \in [0, 1]$ and $u \in S^{n-1}$, so $deg(\theta_R) = deg(g - \overline{p}, B_R(0), 0)$, and the proof is complete.

In light of the successive extensions obtained for Theorem 6.5, we may discuss reasonable ways of eliminating the assumption on the radial limits of g. Let us start, for instance, with the aforementioned result from [68], which states that (6.30) is sufficient for proving the existence of solutions. From a topological point of view, this condition says, in fact, two different things: on the one hand, it says that g does not vanish outside a compact set; on the other hand, it says that its Brouwer degree over the interval $(-R, R)$ is different from zero when R is large. Thus, we might conceive that a natural generalization of Lazer's assumption for a system of n equations is

$$g(u) \neq 0 \quad \text{for } |u| \geq R \tag{6.41}$$

and

$$deg(g, B_R(0), 0) \neq 0. \tag{6.42}$$

This possible extension was analyzed by Ortega and Sánchez in [96], where an example was constructed showing that (6.41) and (6.42) were not sufficient to ensure the existence of a solution. Explicitly, for $n = 2$ complex notation was employed to define the functions

$$g_1(z) := e^{i\text{Re}(z)} \frac{z}{\sqrt{1 + |z|^2}}, \qquad g(z) := g_1(z) - \gamma. \tag{6.43}$$

Then it was shown that, for any $\gamma \in (0, 1)$, the problem

$$z'' + g(z) = \lambda \sin t$$

does not have a 2π-periodic solution when λ is large enough. The proof can be sketched as follows: if z_k is a solution for $|\lambda_k| \to \infty$, then define $w_k = z_k + \lambda_k \sin t$, and hence

$$w_k'' + g_1(w_k - \lambda_k \sin t) = \gamma,$$

so

$$\int_0^{2\pi} g_1(w_k - \lambda_k \sin t) \, dt = 2\pi\gamma.$$

A contradiction follows then from a lemma on oscillating integrals, which states that the integral term in the last equality tends to 0 as $\lambda_k \to \infty$.

The conclusion in [96] was that the existence of radial limits was, in some sense, necessary. However, for the scalar problem we saw that (6.41)–(6.42) could be weakened to a nonasymptotic condition; thus, it is worth attempting to search at once for an extension of (6.32). This work was done by Ruiz and Ward [102], who proved a result that, adapted to our situation, reads as follows.

Theorem 6.11. *(Ruiz–Ward) Let $p \in C([0,1])$, and let $g \in C(\mathbb{R}^n, \mathbb{R}^n)$ be bounded. Further, assume there exists a bounded open $D \subset \mathbb{R}^n$ such that:*

1.
$$\overline{p} \notin co(g(\overline{B_r(v)})) \qquad for\ each\ v \in \partial D \tag{6.44}$$

where $r = \frac{\|p\|_\infty + \|g\|_\infty}{4\pi}$;

2. $deg(g - \overline{p}, D, 0) \neq 0$.

Then problem (6.40) has at least one solution.

An equivalent formulation provides a better understanding of the geometric nature of (6.44): for each $v \in \partial D$ there exists a hyperplane H_v passing through the origin such that $g(B_r(v)) \subset \mathbb{R}^n \setminus (\overline{p} + H_v)$. The following figure, taken from [6], illustrates the condition when $\overline{p} = 0$:

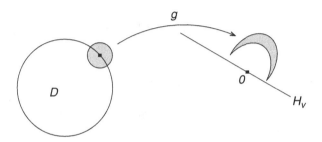

It is an easy exercise to prove that the Nirenberg condition implies (6.44) for $D = B_R(0)$ with R large enough (Exercise 6.5). However, unlike in Theorem 6.10, the result by Ruiz and Ward allows g to have no radial limits and even rotate around $\overline{p}$, although not too fast. We conclude that (6.44) constitutes an accurate extension of (6.32). The presence of the convex hull should not be a surprise when one is dealing with systems, especially after noticing the trivial fact that when $n = 1$, "convex" and "connected" are just synonyms.

The proof of Theorem 6.11 follows the same scheme as the previous ones. The domain for the continuation is chosen again as in (6.33), but now everyone lives in $\mathbb{R}^n$. As happened in the case $n = 1$, it is seen that if u is a solution of (6.20), then $\|u - \overline{u}\|_\infty < r$, that is, $u(t) \in B_r(\overline{u})$ for all t.

If we suppose that $\overline{u} \in \partial D$, then from (6.44) we deduce that $\overline{p} \notin co(g(Im(u)))$. Hence, we may apply the geometric version of the Hahn–Banach theorem, the one that establishes the existence of a hyperplane H that separates $\overline{p}$ and $co(g(Im(u)))$ or, equivalently, of a vector $w \in \mathbb{R}^n$ such that $g(u(t)) \cdot w > \overline{p} \cdot w$ for all t. Either interpretation allows us to derive a contradiction from the equality $\int_0^1 g(u(t)) dt = \overline{p}$ in the following way: just multiply both sides by w to obtain

$$\overline{p} \cdot w = \int_0^1 g(u(t)) \, dt \cdot w = \int_0^1 g(u(t)) \cdot w \, dt > \overline{p} \cdot w.$$

Remark 6.6. An analogous conclusion is obtained directly from the mean value theorem for vector-valued integrals, which is proven precisely by the Hahn–Banach theorem: if $\gamma : [0,1] \to \mathbb{R}^n$ is continuous, then $\overline{\gamma} \in co(Im(\gamma))$. In our previous proof, if one considered $\gamma(t) = g(u(t))$, then $\overline{p} = \int_0^1 \gamma(t) \, dt \in co(Im(\gamma))$, so it is seen that $\overline{u} \notin \partial D$.

Remark 6.7. As seen in Sect. 6.3.1, the boundedness condition on g can be relaxed in different situations. The reader may easily obtain a value of r (depending also on D) when g is sublinear or satisfies an analog of condition (6.38). Moreover, if $g = \nabla G$ for some C^1 function $G : \mathbb{R}^n \to \mathbb{R}$, then the value $r = \frac{1}{2|a|} \|p\|_{L^2}$ is obtained in exactly the same way as before when a nonzero term au' (or $a \cdot u$ for some $a \in \mathbb{R}^n$) is added to (6.40). Finally, the analog of the one-side bounded nonlinearity for a system of equations is

$$Im(g) \subset \xi + \left(\mathbb{R}^n \backslash \bigcup_{j=1}^n H_j \right), \tag{6.45}$$

where $\xi \in \mathbb{R}^n$, and $H_j \subset \mathbb{R}^n$ are linearly independent hyperplanes through the origin. In other words, (6.45) says that the image of g is contained in an *angular sector* of $\mathbb{R}^n$ (Exercise 6.7).

The last result sheds some light on the example in [96], with $\overline{p} = 0$. The function g in (6.43) satisfies (6.41) but not (6.44), and the reason is easy to understand: the value of r gets larger as the value of $\|p\|_\infty$ increases. In particular, if $r > \pi$ and $D = B_R(0)$ with $R \gg 0$, then for $|z| = R$ the curve $g_1(z+t)$ with $t \in (-r, r)$ remains always near the circumference of radius 1, whereas it contains antipodal points in each direction. The difficulty disappears if one considers, more generally,

$$g_\rho(z) := e^{i \frac{\mathrm{Re}(z)}{\rho}} \frac{z}{\sqrt{1 + |z|^2}}$$

for some large enough $\rho > 0$. Note that $\|g_\rho\|_\infty$ does not depend on ρ, so once p is fixed, we may assume that r is fixed, too. Hence, condition (6.44) is satisfied when ρ is sufficiently large: an approximate lower bound for $\rho = \rho(p)$ would be $\frac{2r}{\pi}$ since in this case the curve $g_\rho(z+t)$ with $t \in (-r, r)$ does not contain antipodal points. But condition (6.44) fails for those values of ρ that are smaller than $\rho(p)$, that is, for those nonlinearities g_ρ that rotate too fast around the origin.

Note also that the effect of rotation is produced only when the image of the *whole* ball $B_r(z)$ is considered: if we focus on the vertical strip

$$\mathscr{S}(z) := \{ u \in B_r(z) : |\mathrm{Re}(u) - \mathrm{Re}(z)| < \delta \},$$

then we observe that its image under g remains in the same half-plane when δ is small. This motivates a refinement of Theorem 6.11, inspired in previous results on *rapid oscillations* for the case $n = 1$ (see, e.g., [54, 55] and references therein).

For example, assume that $\overline{p} = 0$ and that g is bounded from below with $g < 0$ over some interval $(a - r, a + r)$, but no interval of positivity of g is long enough to satisfy (6.32). In such a sad situation, it is possible to compensate this "smallness" by assuming that g is larger than an appropriate constant C over some convenient subset $(b - \delta, b + \delta)$ of one of these positivity intervals. The value of C shall depend not only on p but also on the length of the positivity interval: faster oscillations require larger values of g. Note that, as seen in Sect. 6.3.1, the value of r is obtained using only the lower bound of g. This means that g is not necessarily bounded from above; in particular, if, for example, g is negative for $u \ll 0$ and behaves as

$$u^2[\sin u^2 + 1] + \sin u^2$$

for $u \gg 0$, then it is proven that, even if the length of the positivity intervals of g tends to 0 as $u \to +\infty$, the large factor u^2 guarantees solvability for any p.

The main theorem in [6] can be regarded as an extension of this idea (perhaps still obscure) to the case $n > 1$. The generalization is rather straightforward for the case of *weakly coupled* systems: roughly speaking, systems that asymptotically behave as an uncoupled system. But the issue is more difficult in the general case of *rapidly rotating* nonlinearities.

For simplicity, henceforth we shall consider only the case $\overline{p} = 0$ and, moreover, that g is bounded. The goal is to prove the existence of solutions of (6.10) in situations where condition (6.44) is violated. More precisely, according to its geometric interpretation, we shall allow $g(B_r(v))$ to intersect the hyperplane H_v, provided that g maps an appropriate subset of $B_r(v)$ sufficiently far from H_v. This subset shall have the form of a strip of width 2δ, namely,

$$\mathscr{S}(v) = \{u \in B_r(v) : |(u - v) \cdot \xi_v| < \delta\},$$

for some $\xi_v \in S^{n-1}$ and $\delta > 0$. Then we have the following theorem.

Theorem 6.12. *Let $g \in C(\mathbb{R}^n, \mathbb{R}^n)$ be bounded, let $p \in C([0,1])$ such that $\overline{p} = 0$, and let $r = \frac{\|p\|_\infty + \|g\|_\infty}{4\pi}$. Further, assume there exists $D \subset \mathbb{R}^n$ open and bounded with the following properties:*

1. For every $v \in \partial D$ there exist a hyperplane H_v passing through the origin and a strip $\mathscr{S}(v)$ of width 2δ such that $g(\mathscr{S}(v)) \subset \mathbb{R}^n \backslash H_v$ and

$$\mathrm{dist}(g(\mathscr{S}(v)), H_v) > \left(\frac{r}{\delta} - 1\right) \mathrm{dist}(g(u), H_v) \tag{6.46}$$

for every $u \in B_r(v)$ with $g(u) \in H_v^-$, where H_v^- denotes the closure of the connected component of $\mathbb{R}^n \backslash H_v$ not containing $g(\mathscr{S}(v))$;
2. $\deg(g, D, 0) \neq 0$.

Then there exists at least one solution u of (6.10) such that $\overline{u} \in D$ and $\|u - \overline{u}\|_\infty < r$.

The difference between condition (6.44) and condition (6.46) is shown in the following figure:

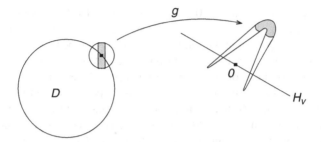

Condition (6.44) requires that the image under g of the whole ball $\overline{B}_r(v)$ lies on one side of a hyperplane H_v passing through the origin, whereas condition (6.46) only requires the image of some strip $\mathscr{S}(v)$ to lie on one side of H_v. Nevertheless, the image of the rest of the ball may cross the hyperplane, thus allowing for fast rotations of g. In particular, for the previous example with g_ρ, Theorem 6.12 provides, for a given p, solutions for a range of values of ρ that is larger than that established from condition (6.44).

As first glance, the boundedness condition enters into conflict with the fact that condition (6.46) needs large values of g when δ is small; in this sense, *arbitrarily fast rotations are still forbidden*. However, more general versions of the result are possible, assuming that g, instead of being bounded, satisfies one of those conditions mentioned in Remark 6.7 that guarantee the existence of r (see [6] for details).

At this stage, the proof of Theorem 6.12 just requires verifying that if

$$u''(t) = \lambda\,(p(t) - g(u(t))), \qquad u(0) = u(1)\ u'(0) = u'(1)$$

for some $\lambda \in (0,1)$, then $\overline{u} \notin \partial D$. We already know that $\|u - \overline{u}\|_\infty \leq \frac{1}{2}\|u'\|_\infty < r$.

Let w_v be the unit normal vector of H_v such that $g(v) \cdot w_v > 0$; then it is readily seen that condition (6.46) is equivalent to the following one:

For every $v \in \partial D$ there exist a vector $w_v \in S^{n-1}$ and a strip $\mathscr{S}(v)$ of width 2δ such that

$$\inf_{z \in \mathscr{S}(v)} g(z) \cdot w_v + \left(\frac{r}{\delta} - 1\right) g(u) \cdot w_v > 0 \tag{6.47}$$

for every $u \in B_r(v)$ satisfying $g(u) \cdot w_v \leq 0$.

Now suppose $\overline{u} \in \partial D$, and take $w_{\overline{u}} \in S^{n-1}$ and a strip $\mathscr{S}(\overline{u})$ such that (6.47) is fulfilled. Let $K = \min_{t \in [0,1]} g(u(t)) \cdot w_{\overline{u}}$; then, from the homotopy equation we obtain

$$0 = \int_0^1 g(u(t)) \cdot w_{\overline{u}}\,dt = \int_0^1 (g(u(t)) - Kw_{\overline{u}}) \cdot w_{\overline{u}}\,dt + K.$$

In particular, this implies that $K \leq 0$.

Next, consider the function $\varphi(u) := u \cdot \xi_{\overline{u}}$. From the mean value theorem for vector-valued integrals (Remark 6.6), we know that $\overline{u} \in co(Im(u))$, and hence $\varphi(\overline{u}) \in \varphi(Im(u))$. Thus, setting $\overline{t} \in [0,T]$ such that $\varphi(u(\overline{t})) = \varphi(\overline{u})$, we obtain

$$|(u(t) - \bar{u}) \cdot \xi_{\bar{u}}| = |\varphi(u(t)) - \varphi(\bar{u})| \leq |u(t) - u(\bar{t})| = |u'(\eta)(t - \bar{t})| \leq 2r|t - \bar{t}|.$$

We conclude that $u(t) \in \mathscr{S}(\bar{u})$ for $|t - \bar{t}| < \frac{\delta}{2r}$, and, from the periodicity of u, it follows that meas$(A) \geq \frac{\delta}{r}$, where $A = \{t \in [0,1] : u(t) \in \mathscr{S}(\bar{u})\}$ and *meas*(A) denotes its Lebesgue measure. Fix $t_0 \in [0,1]$ such that $g(u(t_0)) \cdot w_{\bar{u}} = K \leq 0$; then

$$0 = \int_0^1 (g(u(t)) - Kw_{\bar{u}}) \cdot w_{\bar{u}} \, dt + K \geq \int_A (g(u(t)) - Kw_{\bar{u}}) \cdot w_{\bar{u}} \, dt + K$$

$$= \int_A (g(u(t)) \cdot w_{\bar{u}} \, dt - K\text{meas}(A) + K \geq \frac{\delta}{r} \inf_{v \in \mathscr{S}(\bar{u})} g(v) \cdot w_{\bar{u}} + \left(1 - \frac{\delta}{r}\right) K$$

$$= \frac{\delta}{r} \left[\inf_{v \in \mathscr{S}(\bar{u})} g(u) \cdot w_{\bar{u}} + \left(\frac{r}{\delta} - 1\right) g(u(t_0)) \cdot w_{\bar{u}}\right].$$

This contradicts (6.47), so the proof is complete.

We end this section with another example in which the geometry plays a crucial role. To begin, recall a condition that was employed many times in previous chapters:

$$f(t,u) \cdot u > 0 \qquad \text{for } |u| = R. \tag{6.48}$$

For the second-order problem $u''(t) = f(t, u(t))$ under the Dirichlet conditions $u(0) = u_0, u(1) = u_1$ with $|u_0|, |u_1| \leq R$, or under periodic conditions, the assumption is sufficient to prove the existence of solutions. Under the scope of the continuation methods, (6.48) is used to prove that solutions of the problem with $0 \leq \lambda \leq 1$ $u''(t) = \lambda f(t, u(t))$ cannot be tangent from inside to $\partial B_R(0)$ at any point $t_0 \in (0,1)$. This simply follows from the fact that, in such a case, the second derivative of the function $\varphi(t) := |u(t)|^2$ should be nonpositive at t_0.

The situation is different when f depends also on u' since the continuation theorem must now be applied, for example, on $C^1([0,1])$, and consequently, bounds for the derivative are required. For this reason, a Nagumo-type condition is usually assumed; an example of this is given in Exercise 6.8. The natural extension of condition (6.48) is

$$f(t,u,v) \cdot u > 0 \qquad \text{for } |u| = R. \tag{6.49}$$

However, one immediately notices that the use we make of the condition does not require its validity for *all* values of v. Indeed, if $u(t) \in \overline{B_R(0)}$ and $|u(t_0)| = R$, then t_0 is a critical point of the previous function φ, that is, $u(t_0) \cdot u'(t_0) = 0$. Thus, we only need to assume that (6.49) holds when $v \perp u$.

Note that (6.49) says that f points outwards from the ball $B_R(0)$ when $|u| = R$, but when one computes the second derivative of φ, namely,

$$\varphi''(t) = 2\left(u(t) \cdot u''(t) + |u'(t)|^2,\right),$$

one very soon realizes that the condition can be weakened. Indeed, f is allowed to point inwards although not too much when the velocity is small. This is the idea behind the condition introduced by Hartman [45] in 1960:

$$f(t,u,v) \cdot u + |v|^2 > 0 \qquad \text{for } |u| = R, \ v \perp u. \tag{6.50}$$

Together with an appropriate Nagumo-type condition, (6.50) allows us to prove the existence of solutions for the Dirichlet problem. A similar result holds for the periodic case, which was first treated in [56].

There are many other extensions of the original result, for different boundary conditions and for more general operators. However, fewer generalizations are known in which the ball $B_R(0)$ is replaced by an arbitrary bounded domain $D \subset \mathbb{R}^n$.

In view of its geometric interpretation, it is not difficult to prove the existence of solutions using (besides the Nagumo-like condition) an analog of (6.49) when D is convex and $u_0, u_1 \in D$. Indeed, it suffices to change (6.49) by

$$f(t,u,v) \cdot v_u > 0 \qquad \text{for } u \in \partial D, \ v \perp v_u,$$

where v_u denotes an outer normal (not necessarily unique) of ∂D at u. For the periodic problem, this fact was observed already in [15] by Bebernes and Schmitt, who assumed, instead of the Nagumo condition, that f had some specific subquadratic growth on v.

In [9], the previous results are extended for a more general (not necessarily convex) bounded domain $D \subset \mathbb{R}^n$. Some results in this direction had been given previously in [41] and [75] (see also [42] and the survey [76]), where the concept of *curvature bound set* is introduced to guarantee that solutions starting inside an appropriate domain will always remain there. An approximate picture of this idea is taht at any point of the boundary of such a set D there exists a smooth tangent surface that allows one to measure the curvature of the solutions touching ∂D from inside.

What is shown in [9] is that, in some sense, if D has a C^2 boundary, then the role of the previously mentioned surfaces is assumed by ∂D itself. Thus, a precise geometric condition generalizing (6.50) is

$$f(t,u,v) \cdot v_u > II_u(v) \text{ for all } (u,v) \in T\partial D. \tag{6.51}$$

Here $T\partial D$ denotes the tangent vector bundle identified, as usual, with a subset of $\mathbb{R}^n \times \mathbb{R}^n$, and $II_u(v)$ is the second fundamental form of the surface (e.g., [31]). This condition requires f to point outside D as much as ∂D is "bent outside" in the direction of the velocity. In particular, when $D = B(0,R)$, the curvature is constantly $\frac{1}{R^{n-1}}$; moreover, direct computation shows that $II_x(y) = -\frac{1}{R}|y|^2$ and $v_u = \frac{u}{R}$, so the new Hartman condition (6.51) takes the form of the original one. Among other results, the existence of solutions under this condition was proven in [9] assuming also

$$|f(t,u,v)| \leq c|v|^2 + K \tag{6.52}$$

for every $u \in \overline{D}$ and some constants c and K with $cR_D < 1$, where R_D denotes the radius of D, defined as the radius of the smallest ball containing D. We shall not give a detailed proof here. The smallness condition $cR_D < 1$ may look strange, but its necessity can be partially understood from the computations in Exercise 6.8.

Also, it is interesting to observe that, although the geometry of D can be much richer than in [45], the assumptions in [9] imply that we are not completely free to choose it. Indeed, it is proven that D must be *contractible*; that is, the identity function $i : D \to D$ is homotopic to a constant. In particular, the results cannot be applied if D has a hole, for example $D = B(0,R) \setminus B(0,r)$, for some $r < R$. This case is interesting since it may be used for *singular problems* in which f is not defined for $u = 0$; an example of such a problem is given in Exercise 6.9. Nonetheless, as mentioned in [9], a sharpened version of the approach could still be used in this case since (6.52) provides bounds for u', and thus (6.51) does not need to be assumed for all v. However, after all the work we've done, this specific task can be left for some other time.

Problems

6.1. Let $f : [0,1] \times \mathbb{R}^n \to \mathbb{R}^n$ be continuous. Using the Leray–Schauder degree, prove that the equation $u''(t) = f(t, u(t))$ with the Dirichlet condition $u(0) = u_0, u(1) = u_1$ has at least one solution in the following cases:

1. f sublinear in u;
2. f nondecreasing in u;
3. $f(t,u) \cdot u \geq 0$ for $|u| = R \geq |u_0|, |u_1|$.

Generalize for $f(t, u(t), u'(t))$ with an appropriate growth condition.

6.2. Let $a : \mathbb{R} \to \mathbb{R}_{>0}$ and $g : \mathbb{R}^3 \to \mathbb{R}_{>0}$ be continuous and T-periodic in t. Assume that g is strictly decreasing in the last two coordinates, and let $\phi(u) := \int_0^T g(t, u, u) \, dt$.

1. Prove that $\phi : \mathbb{R} \to Im(\phi)$ is a homeomorphism.
2. Prove that if $\int_0^T a(t) \, dt \in Im(\phi)$, then the delay differential equation

$$u'(t) = -a(t) + g(t, u(t), u(t - \tau))$$

has at least one T-periodic solution for any $\tau > 0$. Is the condition also necessary?

6.3. (Adapted from [7]) Consider the Nicholson blowfly model with harvesting term

$$x'(t) = -\delta(t)x(t) + P(t)x(t - \tau)e^{x(t - \tau)} - H(t, x(t)),$$

where $\delta, P : \mathbb{R} \to \mathbb{R}_{>0}$ are continuous and T-periodic, $H : \mathbb{R} \times [0, +\infty) \to [0, +\infty)$ is continuous, C^1 in x, and T-periodic in t, and τ is a positive constant. Prove that the condition

$$\delta(t) + \frac{\partial H}{\partial x}(t, 0) < P(t) \text{ for all } t$$

is sufficient for the existence of positive T-periodic solutions.

Hint: apply Theorem 6.3 with $\Omega = \{x \in C_T : m < x(t) < M\}$ for some positive $m \ll 1$ and $M \gg 1$.

6.4. Generalize the Landesman–Lazer theorem to:

1. Other boundary conditions;
2. Higher-order problems.

6.5. 1. Prove the following extension of Nirenberg's theorem: let $g \in C(\mathbb{R}^n, \mathbb{R}^n)$ be bounded, let $p \in C([0,1], \mathbb{R}^n)$, and assume that

    a. There exists a family $\{U_j\}_{j=1,\dots,K}$ of open subsets of S^{n-1} and vectors $w_j \in S^{n-1}$ such that $S^{n-1} \subset \bigcup_{j=1}^{K} U_j$ and the upper limit

$$\limsup_{r \to +\infty} \langle g(ru) - \bar{p}, w_j \rangle$$

    is uniform and strictly negative for $u \in U_j$;

    b. $\deg(g - \bar{p}, B_R(0), 0) \neq 0$ for R large enough.

    Then the periodic problem for system $u''(t) + g(u(t)) = p(t)$ has at least one solution.

2. Prove that the preceding result (and, consequently, the Nirenberg theorem) can be deduced directly from Theorem 6.11.

6.6. Prove the Frederickson–Lazer theorem [37]: let $f : \mathbb{R} \to \mathbb{R}$ be continuous and $p : \mathbb{R} \to \mathbb{R}$ be continuous and 2π-periodic. Let $F(u) := \int_0^u f(s)\,ds$, and assume that

$$-\infty < F(-\infty) < F(u) < F(+\infty) < +\infty$$

for all x. Then the condition

$$\frac{2}{\pi}(F(+\infty) - F(-\infty)) > \sqrt{\alpha_1(p)^2 + \beta_1(p)^2},$$

with α_1 and β_1 as in Sect. 6.1.3, is both necessary and sufficient for the existence of 2π-periodic solutions of the equation

$$u''(t) + f(u(t))u'(t) + u(t) = p(t).$$

Compare with Example 4.5.

    Hint: a continuation argument or the shooting method may be used.

6.7. Let $g \in C(\mathbb{R}^n, \mathbb{R}^n)$ satisfy (6.45), and let u be a solution of

$$u''(t) = \lambda(p(t) - g(u(t))), \qquad u(0) = u(1), u'(0) = u'(1)$$

for some $p \in C([0,1])$ and $\lambda \in (0,1)$. Find an a priori bound for $\|u'\|_\infty$.

6.8. (*Hartman–Nagumo condition*) Let $f : [0,1] \times \mathbb{R}^{2n} \to \mathbb{R}^n$ be continuous, and assume that

$$f(t,u,v) \cdot u + |v|^2 \geq 0 \qquad \text{for all } u, v \in \mathbb{R}^n \text{ such that } |u| = R \text{ and } v \perp u$$

and

$$|f(t,u,v)| \le c\left(u \cdot f(t,u,v) + |v|^2\right) + K$$

for $|u| \le R$ and some constants c, K with $cR < 1$. Prove that the Dirichlet problem

$$\begin{cases} u''(t) = f(t,u(t),u'(t)), \\ u(0) = u_0, \ u(1) = u_1, \end{cases}$$

with $|u_0|, |u_1| \le R$, has at least one solution. Deduce an analogous result for periodic and Neumann conditions.

Hint: find bounds for $u'(t)$ using the identity

$$u'(t) = u(1) - u(0) + \int_0^t su''(s)\,ds - \int_t^1 (1-s)u''(s)\,ds.$$

6.9. * (Adapted from [12]). Let $G : \mathbb{R}^n \backslash \{0\} \to \mathbb{R}$ be a C^1 function, let $p : [0,1] \to \mathbb{R}^n$ be continuous with zero average, and let $a \ne 0$. Assume:

1. There exists $R_0 > 0$ such that $\nabla G(u) \cdot u < p(t) \cdot u$ for all $t \in [0,1]$ and all $u \in \mathbb{R}^n$ with $|u| = R_0$,
2. There exist $R_1 > R_0$ and $w \in \mathbb{R}^n$ such that $\nabla G(u) \cdot w > 0$ for all $u \in \mathbb{R}^n$ with $|u| \ge R_1$.
3. $\lim_{u \to 0} G(u) = +\infty$.

Then the problem

$$u''(t) + au'(t) + \nabla G(u(t)) = p(t), \qquad u(0) = u(1), u'(0) = u'(1)$$

admits at least one solution u such that $\|u\|_\infty \ge R_0$.

Hint: for $n > \frac{1}{R_0}$, set $G_n(u) := \rho\,(n|u|)\,G(u)$, where $\rho : [0,+\infty) \to [0,1]$ is smooth such that $\rho(s) = 0$ for $s \le \frac{1}{2}$ and $\rho(s) = 1$ for $s \ge 1$. Using the continuation method, prove that the problem for G_n has a solution u_n with $R_0 < \|u_n\|_\infty < R_1 + \frac{\|p\|_{L^2}}{|a|}$ and $\|u_n'\|_{L^2} < \frac{\|p\|_{L^2}}{|a|}$. Deduce that a subsequence of u_n converges uniformly to some u, and prove that u is a solution.

Appendix A
Basic Facts on Metric and Normed Spaces

We shall assume that the reader is familiar with the usual topology of $\mathbb{R}^n$ and some well-known concepts such as:

1. Open and closed subsets, closure, interior, density;
2. Limit of sequences and functions, continuity;
3. Compactness;
4. Completeness;
5. Connectedness.

All these ideas are extended in a straightforward manner to a general metric space, that is, a set X equipped with a distance function or *metric*) $d : X \times X \to \mathbb{R}_{\geq 0}$ satisfying

1. $d(x,y) = 0 \iff x = y$;
2. $d(x,y) = d(y,x)$ for all $x, y \in X$;
3. $d(x,y) \leq d(x,z) + d(z,y)$ for all $x, y, z \in X$.

In most cases we shall work in *normed* spaces, that is, a vector space X equipped with a norm $\| \cdot \| : X \to \mathbb{R}_{\geq 0}$ such that

1. $\|x\| = 0 \iff x = 0$;
2. $\|ax\| = |a| \|x\|$ for all $x \in X$ and $a \in \mathbb{R}$;
3. $\|x + y\| \leq \|x\| + \|y\|$ for all $x, y \in X$.

This induces a metric given by $d(x,y) := \|x - y\|$. If X is complete for this metric, then it is called a *Banach space*.

Our typical examples of Banach spaces shall be

1. $\mathbb{R}^n$, with the euclidean norm $|\cdot|$;
2. $C(\overline{\Omega}) := \{f : \overline{\Omega} \to \mathbb{R}^n \text{ continuous}\}$, where $\Omega \subset \mathbb{R}^k$ is open and bounded, with the norm $\|f\|_\infty := \max_{x \in \overline{\Omega}} |f(x)|$;
3. $C^j(\overline{\Omega}) := \{f : \overline{\Omega} \to \mathbb{R}^n \text{ of class } C^j\}$, where $\Omega \subset \mathbb{R}^k$ is open and bounded, with the norm $\|f\|_{C^j} := \max\{\|f\|_\infty, \|Df\|_\infty, \ldots, \|D^j f\|_\infty\}$; by C^j function on $\overline{\Omega}$ we understand a function that can be extended to a C^j function defined in a neighborhood of $\overline{\Omega}$.

P. Amster, *Topological Methods in the Study of Boundary Value Problems*, Universitext, DOI 10.1007/978-1-4614-8893-4, © Springer Science+Business Media New York 2014

The two following classical results are very frequently used throughout the book.

Theorem A.1. *(Stone–Weierstrass) Let $\Omega \subset \mathbb{R}^k$ be open and bounded. Then the set of all the polynomials $p : \overline{\Omega} \to \mathbb{R}^k$ is dense in $C(\overline{\Omega})$.*

Theorem A.2. *(Arzelá–Ascoli) Let $\Omega \subset \mathbb{R}^k$ be open and bounded, and let $A \subset C(\overline{\Omega})$ be bounded, that is, there exists $R \geq 0$ such that $\|f\|_\infty \leq R$ for all $f \in A$. Further, assume that A is equicontinuous, that is, for all $\varepsilon > 0$ there exists $\delta > 0$ such that*

$$|x - y| < \delta \Rightarrow |f(x) - f(y)| < \varepsilon \qquad \text{for all } f \in A.$$

Then $\overline{A}$ is compact.

Sometimes we shall need more structure, so a vector space X shall be endowed with an *inner product*, i.e., a function $\langle \cdot, \cdot \rangle : X \times X \to \mathbb{R}$ satisfying

1. $\langle x, x \rangle \geq 0$ for all $x \in X$, with equality if and only if $x = 0$;
2. $\langle x, y \rangle = \langle y, x \rangle$ for all $x, y \in X$;
3. $\langle ax + y, z \rangle = a\langle x, z \rangle + \langle y, z \rangle$ for all $x, y, z \in X$ and $a \in \mathbb{R}$.

If we define $\|x\| := \sqrt{\langle x, x \rangle}$, then the *Cauchy–Schwarz* inequality

$$|\langle x, y \rangle| \leq \|x\| \|y\|$$

is easily verified. In particular, this allows us to prove that $\|\cdot\|$ defines a norm. If X is complete for this norm, then it is called a *Hilbert space*. With the obvious exception of $\mathbb{R}^n$, Hilbert spaces are very rarely used in this book. The Lebesgue space $L^2(a, b)$ is mentioned on several occasions, but as was stated in the introduction, measure theory is not really employed. In most cases, our "use" of L^2 is limited to the idea of having an inner product $\langle f, g \rangle$ that, in the case of continuous functions f, g, simply reduces to the standard Riemann integral $\int_a^b f(t)g(t)\,dt$. Some important theorems of measure theory, such as the Lebesgue dominated convergence theorem, are applied in a few cases, but always the involved functions are continuous, so the concept of *measurable function* does not play any fundamental role. In fact, all we need from the theorem reduces to the following particular situation: if $f_n : [a, b] \to \mathbb{R}$ is continuous, $\|f_n\|_\infty$ is bounded, and $f_n \to f$ pointwise, then $\int_a^b f_n(t)\,dt \to \int_a^b f(t)\,dt$. We need not worry too much about the meaning of the latter integral if f is not continuous. Incidentally (only in a starred remark), we recall the fact that if $f \in L^2(a, b)$, then $\int_a^t f(s)\,ds$ is an absolutely continuous function whose derivative is f (a.e.) and make a brief mention of the Sobolev spaces H^1 or H^2. Also, in a specific exercise we refer to the space l^2 of real sequences $(x_n)_{n \in \mathbb{N}}$ such that $\sum_{n=1}^\infty x_n^2 < \infty$. But, as the reader may imagine, this mention is completely irrelevant to the general goals of the book.

Appendix B
Brief Review of Ordinary Differential Equations

This appendix is devoted to a quick review of the main aspects of ordinary differential equations used in the book. The first section concerns only very well-known facts of the general theory and is presented in the form of exercises. The same criterion was used for the first part of the second section, which is devoted to general results on second-order linear equations. The remaining parts contain some other aspects of second-order equations, also well known but not always covered in introductory texts.

B.1 Basic Theory

1. *Existence and uniqueness*: let $\Omega \subset \mathbb{R} \times \mathbb{R}^n$ be open, let $(t_0, x_0) \in \Omega$, and let $f : \Omega \to \mathbb{R}^n$ be continuous and locally Lipschitz in x. Prove that the problem

$$\begin{cases} x'(t) = f(t, x(t)) \\ x(t_0) = x_0 \end{cases}$$

   admits a (unique) solution $x : (t_0 - \delta, t_0 + \delta) \to \mathbb{R}^n$ for some $\delta > 0$. Give a lower bound for δ.
2. *Gronwall's lemma*: let $u, v : [t_0, t_1] \to \mathbb{R}_{\geq 0}$ be continuous such that

$$u(t) \leq \alpha + \int_{t_0}^{t} u(s)v(s)\, ds$$

   for all t and some $\alpha \geq 0$. Prove that

$$u(t) \leq \alpha e^{\int_{t_0}^{t} v(s)\, ds}.$$

3. a. *Continuous dependence*: let Ω, $(t_0, x_0) \in \Omega$, f and $x(t)$ as in problem 1, and consider $(\tilde{t}_0, \tilde{x}_0) \in B_r(t_0, x_0)$ for some small enough $r > 0$. Prove that the solution $\tilde{x}$ of the problem

$$\begin{cases} \tilde{x}'(t) = f(t, \tilde{x}(t)) \\ \tilde{x}(\tilde{t}_0) = \tilde{x}_0 \end{cases}$$

satisfies

$$|\tilde{x}(t) - x(t)| \le \alpha e^{L|t - t_0|}$$

for some constants $\alpha = \alpha(\tilde{t}_0, \tilde{x}_0), L \ge 0$, and all t close to t_0, with $\alpha(\tilde{t}_0, \tilde{x}_0) \to 0$ as $(\tilde{t}_0, \tilde{x}_0) \to (t_0, x_0)$. Deduce that the flow, defined as $\phi(t, t_0, x_0) := x(t)$, is a continuous function.

b. Prove that if f is C^k with respect to x, then ϕ is C^k with respect to x_0.

4. a. Let Ω, $(t_0, x_0) \in \Omega$, and f be as before, and let $K \subset \Omega$ be compact. Prove that if a solution of the problem $x'(t) = f(t, x(t))$ is defined in $[t_0, t_1)$ and cannot be extended up to t_1, then there exists $\delta > 0$ such that $(t, x(t)) \notin K$ for $t \in (t_1 - \delta, t_1)$. Deduce that if $[t_0, t_1] \times \mathbb{R}^n \subset \Omega$ then $|x(t)| \to \infty$ for $t \to t_1^-$.

b. Prove that if $f : [t_0, t_1] \times \mathbb{R}^n \to \mathbb{R}^n$ is continuous, locally Lipschitz in x, and has linear growth in x (that is, $|f(t, x)| \le a|x| + b$), then any solution of the problem $x'(t) = f(t, x(t))$ can be extended to $[t_0, t_1]$.

c. Let $A : \mathbb{R} \to \mathbb{R}^{n \times n}$ be continuous. Prove that the solutions of the linear problem $x'(t) = A(t)x(t)$ are globally defined in $\mathbb{R}$ and form an n-dimensional linear subspace of $C(\mathbb{R}, \mathbb{R}^{n \times n})$. Deduce an analogous result for a linear scalar equation of order n.

B.2 Second-Order Equation

1. Prove that the problem

$$u''(t) + a(t)u'(t) + b(t)u(t) = \psi(t),$$

with $a, b, \psi : [0, T] \to \mathbb{R}$ continuous, is equivalent to the equation $Lu = \varphi$, where

$$Lu(t) = (-p(t)u'(t))' + q(t)u(t)$$

for some $q, \varphi : [0, T] \to \mathbb{R}$ continuous and $p : [0, T] \to \mathbb{R}_{>0}$ of class C^1.

2. Let L be as in the previous exercise, and let $\{u_1, u_2\}$ be a basis of solutions of the homogeneous problem $Lu = 0$. Prove that $w := p(u_1 u_2' - u_2 u_1')$ is a nonzero constant.

3. Let L, u_1, u_2, and w be as before, and let $\varphi : [0, T] \to \mathbb{R}$ be continuous. Prove that all the solutions of the equation $Lu = \varphi$ can be written as $u = c_1 u_1 + c_2 u_2$, with

$$c_1(t) = k_1 + \frac{1}{w} \int_0^t u_2(s)\varphi(s)\, ds,$$

$$c_2(t) = k_2 - \frac{1}{w} \int_0^t u_1(s)\varphi(s)\, ds.$$

4. Let L be as before. Prove that if the Dirichlet problem

$$Lu = 0, \qquad u(0) = u(T) = 0$$

has only a trivial solution, then the nonhomogeneous problem

$$Lu = \varphi, \qquad u(0) = u_0, \quad u(T) = u_T$$

has a unique solution for any continuous φ and boundary data $u_0, u_T \in \mathbb{R}$. Obtain similar conclusions for the Neumann and periodic conditions.

5. Let L be as before, with $q \geq 0$. Prove that for any continuous φ the problem

$$Lu = \varphi, \qquad u(0) = u_0, \quad u(T) = u_T$$

has a unique solution.

Prove, furthermore, that $L^{-1} : C([0,T]) \to \{u \in C([0,T] : u(0) = u(T) = 0\}$ is continuous, that is, if $\varphi_n \to 0$ uniformly and u_n is the corresponding solution, then $u_n \to 0$ uniformly. Extend the result for Neumann and periodic conditions, assuming also that $q \not\equiv 0$.

B.2.1 Green's Function

In this section we shall see that, in many cases, the unique solution of a linear problem $Lu = \varphi$ can be expressed in terms of an integral operator. Let us start with an intuitive approach to the elementary case

$$\begin{cases} u''(t) = \varphi(t), \\ u(0) = u(T) = 0. \end{cases}$$

Simple integration yields

$$u'(t) = c + \int_0^t \varphi(s)\,ds = c + \int_0^T \chi_{[0,t)}(s)\varphi(s)\,ds,$$

where χ_A denotes the characteristic or indicator function of set A. Hence

$$u(t) = \int_0^t \left(c + \int_0^T \chi_{[0,r)}(s)\varphi(s)\,ds \right) dr = ct + \int_0^T \chi_{[0,t)}(r) \left(\int_0^T \chi_{[0,r)}(s)\varphi(s)\,ds \right) dr$$

$$= ct + \int_0^T \varphi(s) \left(\int_0^T \chi_{[0,t)}(r)\chi_{[s,T]}(r)\,dr \right) ds = ct + \int_0^T \varphi(s)\psi(t,s)\,ds,$$

where

$$\psi(t,s) := (t-s)\chi_{[0,t)}(s) = \begin{cases} 0 & \text{if } t \leq s, \\ t-s & \text{if } t > s. \end{cases}$$

Moreover, from the condition $u(T) = 0$ we deduce

$$c = -\frac{1}{T}\int_0^T \varphi(s)\psi(T,s)\,ds = -\frac{1}{T}\int_0^T \varphi(s)(T-s)\,ds.$$

Thus,

$$u(t) = \int_0^T G(t,s)\varphi(s)\,ds,$$

where

$$G(t,s) = \begin{cases} -\frac{t}{T}(T-s) & \text{if } t \le s, \\ -\frac{s}{T}(T-t) & \text{if } t > s. \end{cases}$$

The function $G : [0,T]^2 \to \mathbb{R}$ is called *Green's function* associated to the problem.

In particular, the following *Green representation formula* is obtained for arbitrary $w \in C^2([0,T])$:

$$w(t) = \frac{T-t}{T}w(0) + \frac{t}{T}w(T) + \int_0^1 G(t,s)w''(s)\,ds. \tag{B.1}$$

As an exercise, the reader may try to obtain similar expressions for Neumann and periodic conditions and generalize for the operator $L_\lambda u = u'' - \lambda u$. What is the explicit form of G in this case? Are all values of $\lambda \in \mathbb{R}$ allowed? *Hint:* use, for example, the variation of parameters formula obtained in Exercise 3 of the previous section.

The preceding situation can be extended to a more abstract context with the help of some standard results in the theory of Hilbert spaces. We shall restrict ourselves to situations of our interest, although a much more general setting is possible.

Consider the equation $Lu = \varphi$, and assume, for some $X \hookrightarrow C([0,1])$, that the operator $L : X \to L^2(0,1)$ is an isomorphism. Then we may define, for each fixed t, the mapping $T_t : L^2(0,1) \to \mathbb{R}$ given by $T_t\varphi = u(t)$, where u is the unique solution of the problem $Lu = \varphi$. Because T_t is linear and

$$|u(t)| \le \|u\| = \|L^{-1}\varphi\| \le c\|\varphi\|$$

for some constant c, we deduce that T_t is continuous. By Riesz's theorem, T_t can be represented by an element $G_t \in L^2(0,1)$, that is,

$$T_t\varphi = \int_0^1 G_t(s)\varphi(s)\,ds$$

for all $\varphi \in L^2(0,1)$. Define $G(t,s) := G_t(s)$; then

$$u(t) = \int_0^1 G(t,s)\varphi(s)\,ds$$

for all t. It is readily seen that G is an element of $L^2((0,1) \times (0,1))$, with

$$\|G\|_{L^2}^2 = \int_0^1 \int_0^1 G(t,s)^2 \, ds \, dt = \int_0^1 \|G_t\|_{L^2}^2 \, dt \le c^2.$$

B.2.2 Poincaré-Type Inequalities

Throughout the book, the inequality

$$\|u\|_{L^2} \le \frac{1}{\pi} \|u'\|_{L^2},$$

valid for all C^1 functions u such that $u(0) = u(1) = 0$, is used repeatedly. It is not difficult, by simple integration, to deduce the existence of *some* constant c such that $\|u\|_{L^2} \le c\|u'\|_{L^2}$; indeed, one may just write

$$|u(t)| = \left| \int_0^t u'(s) \, ds \right| \le t^{1/2} \left(\int_0^t u'(s)^2 \, ds \right)^{1/2}$$

and conclude that

$$\|u\|_{L^2}^2 \le \int_0^1 t \int_0^t u'(s)^2 \, ds \, dt \le \frac{1}{2} \|u'\|_{L^2}^2.$$

However, very soon one obtains the feeling that the constant $c = \frac{\sqrt{2}}{2}$ thus obtained is not optimal; among other things, observe that the condition $u(1) = 0$ was not used at all.

There is a simple way to show that the value of the constant can be improved up to $c = \frac{1}{\pi}$: for arbitrary u, it suffices to define $\phi(t) = u(t)^2 \cot(\pi t)$ and compute

$$\phi'(t) = 2u(t)u'(t) \cot(\pi t) - \pi u(t)^2 (1 + \cot^2(\pi t))$$

$$= \left(\frac{u'(t)}{\sqrt{\pi}} \right)^2 - \pi u(t)^2 - \left(\sqrt{\pi} u(t) \cot(\pi t) - \frac{u'(t)}{\sqrt{\pi}} \right)^2 \le \frac{u'(t)^2}{\pi} - \pi u(t)^2.$$

Thus,

$$0 = \phi \Big|_0^{1/2} \le \int_0^{1/2} \left[\frac{u'(t)^2}{\pi} - \pi u(t)^2 \right] dt,$$

and hence

$$\int_0^{1/2} u(t)^2 \, dt \le \frac{1}{\pi^2} \int_0^{1/2} u'(t)^2 \, dt.$$

In the same way [here we use that $u(1) = 0$!], it is verified that

$$\int_{1/2}^1 u(t)^2 \, dt \le \frac{1}{\pi^2} \int_{1/2}^1 u'(t)^2 \, dt,$$

and the result follows.

Rescaling the interval, we obtain a more general version, as follows.

Lemma B.1. *(Poincaré Inequality) If u is a C^1 function such that $u(0) = u(T) = 0$,*
then

$$\|u\|_{L^2} \le \frac{T}{\pi} \|u'\|_{L^2}.$$

Similar considerations yield another important inequality.

Lemma B.2. *(Wirtinger Inequality) If u is a C^1 function such that $u(0) = u(T)$, then*

$$\|u - \bar{u}\|_{L^2} \le \frac{T}{2\pi} \|u'\|_{L^2}.$$

In this latter case, Fourier series provide a simple and elegant proof. Assume that
$T = 2\pi$, and write

$$u(t) - \bar{u} = \sum_{n=1}^{\infty} a_n \cos nt + b_n \sin nt;$$

then

$$u'(t) = \sum_{n=1}^{\infty} n(b_n \cos nt - a_n \sin nt),$$

and hence

$$\|u - \bar{u}\|_{L^2}^2 = \sum_{n=1}^{\infty} \pi(a_n^2 + b_n^2) \le \sum_{n=1}^{\infty} \pi n^2(a_n^2 + b_n^2) = \|u'\|_{L^2}^2.$$

Although effective, the preceding arguments are not completely satisfactory for
several reasons. On the one hand, they look like very ingenious tricks, but, besides the computations, they do not provide a convincing explanation of why the
inequality must be true. On the other hand, the arguments are hardly generalized to
other contexts. Finally (on the third hand?), the previous proofs do not ensure that
the values of the Poincaré and Wirtinger constants are optimal.

In what follows, we shall sketch a proof that shows the deep connection between
the best possible constant in the Poincaré inequality and the theory of eigenvalues.
Similar connections can be found for Wirtinger and other famous inequalities. Some
details are beyond the scope of this text, but the basic ideas are simple and easy to
understand.

Let $X = \{u \in C^1([0,1]) : u(0) = u(1) = 0\}$ be equipped with the inner product
$\langle u, v \rangle := \int_0^1 u'(t)v'(t)\, dt$. As mentioned in Example 2.2, the completion of X is a
well-known Hilbert space called $H_0^1(0,1)$; however, nothing of this will really be
needed. Let us define the functionals

$$I(u) = \int_0^1 u'(t)^2\, dt, \qquad J(u) = \int_0^1 u(t)^2\, dt$$

and consider the problem of finding a minimum of I subject to the constraint
$J(u) = 1$. Using standard tools in functional analysis, it is easy to see that an absolute

minimum u is achieved in the completion of X, but, in fact, u is a C^2 function. Moreover, by the theory of Lagrange multipliers (Exercise 2.11), there exists a number λ_1 such that $DI(u) = \lambda_1 DJ(u)$. In other words,

$$\int_0^1 u'(t)\varphi'(t)\,dt = \lambda_1 \int_0^1 u(t)\varphi(t)\,dt$$

for all φ. Integration by parts yields

$$\int_0^1 (u''(t) + \lambda_1 u(t))\varphi(t)\,dt = 0$$

for all φ, so we deduce that

$$u''(t) + \lambda_1 u(t) = 0$$

for all t. Now we are very lucky because this equation is easy to solve. Since $u(0) = 0$, it is seen that $u(t) = a\sin(\sqrt{\lambda_1}t)$ for some constant a, chosen in such a way that $J(u) = \int_0^1 u(t)^2\,dt = 1$. On the other hand, $u(1) = 0$, which implies $\sqrt{\lambda_1} = k\pi$. Finally, observe that

$$I(u) = \int_0^1 u'(t)^2\,dt = -\int_0^1 u''(t)u(t)\,dt = \lambda_1 \int_0^1 u(t)^2\,dt = \lambda_1 = (k\pi)^2.$$

Because u is an absolute minimum, we conclude that $k = 1$. In other words,

$$\pi^2 = \min_{u\in X,\|u\|_{L^2}=1} \int_0^1 u'(t)^2\,dt.$$

Thus, for arbitrary $u \neq 0$ we may define $w = \frac{u}{\|u\|_{L^2}}$, and hence $\int_0^1 w'(t)^2 \geq \pi^2$, that is,

$$\|u'\|_{L^2}^2 \geq \pi^2 \|u\|_{L^2}^2,$$

as we wished to prove.

Many other inequalities are used in different applications. For example, in most cases we employ uniform bounds instead of L^2 bounds: if u is C^1 and $u(0) = u(T)$, then for arbitrary $t_0 < t_1$

$$u(t_1) - u(t_0) = \int_{t_0}^{t_1} u'(t)\,dt \leq \int_{t_0}^{t_1} [u'(t)]^+\,dt \leq \int_0^T [u'(t)]^+\,dt$$

and

$$u(t_1) - u(t_0) = \int_{t_0}^{t_1} u'(t)\,dt \geq -\int_{t_0}^{t_1} [u'(t)]^-\,dt \geq -\int_0^T [u'(t)]^-\,dt,$$

where, as usual, $A^+ := \max\{A,0\}$ and $A^- := \max\{-A,0\}$. Now observe that

$$0 = \int_0^T u'(t)\,dt = \int_0^T ([u'(t)]^+ - [u'(t)]^-)\,dt,$$

so

$$\int_0^T [u'(t)]^+ \, dt = \int_0^T [u'(t)]^- \, dt = \frac{1}{2} \int_0^T \left([u'(t)]^+ + [u'(t)]^- \right) dt = \frac{1}{2} \int_0^T |u'(t)| \, dt.$$

This implies

$$u_{max} - u_{min} \leq \frac{1}{2} \int_0^T |u'(t)| \, dt \leq \frac{\sqrt{T}}{2} \|u'\|_{L^2} \leq \frac{T}{2} \|u'\|_\infty$$

and, in particular (using the mean value theorem),

$$\|u - \bar{u}\|_\infty \leq \frac{1}{2} \int_0^T |u'(t)| \, dt \leq \frac{\sqrt{T}}{2} \|u'\|_{L^2} \leq \frac{T}{2} \|u'\|_\infty.$$

If, moreover, u is C^2, then by Rolle's theorem we may write, for some t_0,

$$u'(t) = \int_{t_0}^t u''(s) \, ds;$$

thus,

$$\|u'\|_\infty \leq \int_0^T |u''(s)| \, ds \leq \sqrt{T} \|u''\|_{L^2} \leq T \|u''\|_\infty$$

and

$$u_{max} - u_{min} \leq \frac{T}{2} \|u'\|_\infty \leq \frac{T^{3/2}}{2} \|u''\|_{L^2} \leq \frac{T^2}{2} \|u''\|_\infty.$$

Finally, when the condition $u'(0) = u'(T)$ is also assumed, one is tempted to apply the same idea twice and rapidly conclude that

$$\|u - \bar{u}\|_\infty \leq u_{max} - u_{min} \leq \frac{T}{2} \|u'\|_\infty \leq \frac{T^2}{4} \|u''\|_\infty.$$

This is certainly true, although in this case the bound may be improved. Indeed, observe that

$$\int_0^T u'(t)^2 \, dt = -\int_0^T u''(t)(u(t) - \bar{u}) \, dt \leq \|u - \bar{u}\|_{L^2} \|u''\|_{L^2},$$

so we deduce

$$\|u'\|_{L^2}^2 \leq \|u - \bar{u}\|_{L^2} \|u''\|_{L^2} \leq \frac{T}{2\pi} \|u'\|_{L^2} \|u''\|_{L^2}.$$

Hence

$$\|u - \bar{u}\|_\infty \leq u_{max} - u_{min} \leq \frac{\sqrt{T}}{2} \|u'\|_{L^2} \leq \frac{T^{3/2}}{4\pi} \|u''\|_{L^2} \leq \frac{T^2}{4\pi} \|u''\|_\infty.$$

If one is dealing, instead, with the Dirichlet condition $u(0) = u(T) = 0$, then the computation is slightly different. Because the boundary terms in the integration by parts need to vanish, we now write

$$\int_0^T u'(t)^2 \, dt = -\int_0^T u''(t)u(t) \, dt \leq \|u\|_{L^2}\|u''\|_{L^2}$$

and use the Poincaré inequality to conclude

$$\|u\|_\infty \leq u_{max} - u_{min} \leq \frac{\sqrt{T}}{2}\|u'\|_{L^2} \leq \frac{T^{3/2}}{2\pi}\|u''\|_{L^2} \leq \frac{T^2}{2\pi}\|u''\|_\infty.$$

However, if only uniform bounds are required, then the latter inequality can be improved as well: just employ Green's representation formula (B.1) to obtain

$$|u(t)| \leq \int_0^T |G(t,s)u''(s)| \, ds \leq \int_0^T |G(t,s)| \, ds \|u''\|_\infty.$$

A simple computation shows that

$$\int_0^T |G(t,s)| \, ds = \frac{1}{T}\left(\int_0^t s(T-t) \, ds + \int_t^T t(T-s) \, ds\right) = \frac{t(T-t)}{4} \leq \frac{T^2}{8},$$

so we conclude

$$\|u\|_\infty \leq \frac{T^2}{8}\|u''\|_\infty.$$

This value is already sharp, as the obvious example $u(t) := t(T-t)$ shows. The reader may verify that the use of (B.1) does not provide a more accurate uniform estimate for $u - \bar{u}$ in the case of periodic conditions.

B.2.3 Maximum Principles

When dealing with the method of upper and lower solutions, we made intensive use of the following comparison principle: if $u, v \in C^2([0,1])$ satisfy

$$u''(t) \leq v''(t) \text{ for all } t, \qquad u(0) \geq v(0), u(1) \geq v(1),$$

then $u(t) \geq v(t)$ for all t. Due to linearity, this is equivalent to the so-called maximum principle.

Maximum principle: if $u \in C^2([0,1])$ satisfies

$$u''(t) \leq 0 \text{ for all } t, \qquad u(0) \geq 0, u(1) \geq 0,$$

then $u(t) \geq 0$ for all t.

There are several ways of proving this elementary result, for instance:

1. Assume firstly that $u''(t) < 0$ for all t, and let t_0 be the point where the absolute minimum of u is achieved. If $u(t_0) < 0$, then $t_0 \in (0, 1)$ and $u''(t_0) < 0$, a contradiction. For the general case, set $u_n(t) := u(t) + \frac{t(1-t)}{n}$; then $u_n(0) = u(0) \geq 0$, $u_n(1) = u(1) \geq 0$, and $u_n''(t) = u''(t) - \frac{2}{n} < 0$ for all t. It follows that $u_n \geq 0$ for all n, and taking limit we deduce that $u \geq 0$.

2. Write $u^- = \max\{-u, 0\}$. Since $u''(t)u^-(t) \leq 0$ for all t, integration by parts yields

$$\int_0^1 [u^-]'(t)^2 \, dt = -\int_0^1 u''(t)u^-(t) \, dt \leq 0.$$

This implies that $u^- \equiv 0$ and, thus, $u \geq 0$.

3. Use the Green representation formula (B.1):

$$u(t) = (1-t)u(0) + tu(1) + \int_0^1 G(t, s)u''(s) \, ds.$$

Because $u(0), u(1) \geq 0$, the first two terms are nonnegative; moreover, the integral term is also nonnegative since G and u'' are nonpositive.

4. Simply observe that u is concave over the interval $[0, 1]$ and nonnegative on its boundary.

No doubt, the last proof is the easiest one, but the first three have an important advantage: they can be extended to more general situations. In this book, we only make use of the version presented in the following lemma, which is extensible to other boundary conditions. In the same package, we include an existence-uniqueness result, which is proven in Exercise 5 of Sect. B.2.

Lemma B.3. *Let* $a \in C([0, 1])$ *satisfy* $a \geq 0$. *If* $u \in C^2([0, 1])$ *satisfies*

$$u''(t) - a(t)u(t) \leq 0 \text{ for all } t, \qquad u(0), u(1) \geq 0,$$

then $u(t) \geq 0$ *for all* t. *Moreover, for arbitrary* $\varphi \in C([0, 1])$ *the problem*

$$u''(t) - a(t)u(t) = \varphi(t), \qquad u(0) = u(1) = 0$$

has a unique solution.

Remark B.1. The reader may verify that the maximum principle given in Lemma B.3 still holds if $a(t) \geq c > -\pi^2$ for all t.

Hints and Solutions to Selected Exercises

Problems from Chap. 1

1.1 Obtain the following expression for the solution with initial data $u(0) = u_0$ and $u'(0) = \lambda$:

$$u_\lambda(t) = u_0 + \frac{\lambda}{a}\left(1 - e^{-at}\right) + \int_0^t \left(\int_0^s [p(r) - b\sin u_\lambda(r)]e^{a(r-s)}\,dr\right)ds.$$

Thus, $u_R(1) > u_1 > u_{-R}(1)$ for R large enough. The same result holds for general f. Obviously, the periodic problem might not have any solution; for example, it suffices to take p such that $\int_0^1 p(t)\,dt > |b|$.

1.2 By standard results, the function ϕ is well defined and C^1. Consider the flow $\varphi(\lambda,t) := u_\lambda(t)$, and verify that $w_\lambda(t) := \frac{\partial \varphi}{\partial \lambda}(\lambda,t)$ satisfies

$$w_\lambda''(t) + g'(u_\lambda(t))w_\lambda(t) = 0, \qquad w_\lambda(0) = 0,\ w_\lambda'(0) = 1.$$

In particular, for $\lambda = 0$ compute $w_0(t) = \dfrac{\sin\left(\sqrt{g'(0)}\,t\right)}{\sqrt{g'(0)}}$ and deduce the result.

1.3 Define a truncation function relative to α and β.

1.4 For example, if $|f(x) - x| \leq C$ for all x, then define $g(x) = x - f(x)$. Then, by Brouwer's theorem, g has a fixed point in $\overline{B_C(0)}$. Conversely, if $f : B_1(0) \to \overline{B_1(0)}$ is continuous, then define

$$\tilde{f}(x) = \begin{cases} f(x) & |x| \leq 1 \\ f\left(\frac{x}{|x|}\right) & |x| \geq 1 \end{cases}$$

and $g(x) = x - \tilde{f}(x)$. Then $|x - g(x)| \leq 1$, so $\tilde{f}$ has a fixed point, which is also a fixed point of f.

P. Amster, *Topological Methods in the Study of Boundary Value Problems*, Universitext, DOI 10.1007/978-1-4614-8893-4, © Springer Science+Business Media New York 2014

1.5 For example, if $A \subset (0,1)$ has no supremum, then define $r : \overline{B_1(0)} \to \partial B_1(0)$ by

$$r(x) = \begin{cases} \frac{x}{|x|} & \text{if } |x| \text{ is an upper bound of } A, \\ (1,0) & \text{otherwise,} \end{cases}$$

and verify that r is a C^2 retraction.

1.6 Brouwer's theorem is proven in Sect. 1.4. The no-retraction theorem follows trivially from the lemma. If $\phi : \overline{B_1(0)} \to \mathbb{R}^2$ is continuous such that $\phi(x) \cdot x > 0$ for $x \in \partial B_1(0)$, then identify $\mathbb{R}^2$ with $\mathbb{C}$, and define the homotopy $h(t,s) = s\phi(e^{it}) + (1-s)e^{it}$. The Poincaré–Miranda theorem may be proven, for example, adapting Lemma 1.1 to a square.

1.7 Prove that an appropriate truncated problem has a solution u. Verify that the function $\phi(t) := |u(t)|^2$ cannot reach the value R^2.

1.8 As in the one-dimensional case, find a priori bounds and truncate.

1.9 Imitate the approximation argument in Sect. 1.3.1.

1.10 The first part follows from Proposition 1.1. For the second part, one may place the shooting parameter at $t = T$ or, alternatively, define $y(t) = x(T - t)$ and solve the equivalent problem for y.

1.11 Truncate f and g, and use the Poincaré–Miranda theorem.

1.12 Suppose firstly that g is smooth, and consider the initial value problem in polar coordinates

$$u''(t) + u(t) + g(u(t)) = p(t),$$
$$u(0) = r\cos\theta, \quad u'(0) = r\sin\theta.$$

By the method of variation of parameters for the equation $u'' + u = \varphi$, it is seen that

$$u(t) = \left(u(0) - \int_0^t \varphi(s)\sin s\, ds \right)\cos t + \left(u'(0) + \int_0^t \varphi(s)\cos s\, ds \right)\sin t,$$

so for $u = u_{r,\theta}$ we obtain

$$u(t) = r\cos(\theta - t) + \int_0^t [p(s) - g(u(s))]\sin(t - s)\, ds.$$

Thus, if we define the complex function

$$F(r,\theta) = u'(2\pi) - u'(0) + i(u(0) - u(2\pi)),$$

then we deduce

$$F(r,\theta) = \int_0^{2\pi} [p(s) - g(u(s))]e^{is}\, ds.$$

By the previous formula for $u(t)$ and substitution, we may write

$$F(r,\theta) = \int_0^{2\pi} [p(s) - g(r\cos(\theta - s) + \xi(s))]e^{is}\, ds$$

$$= A + iB - e^{i\theta}\int_0^{2\pi} g(r\cos t + \xi(t))e^{it}\, dt,$$

where $|\xi(t)| \le \|p\|_\infty + \|g\|_\infty$ for all t. Let us compute the limit of this expression as $r \to +\infty$. By periodicity, we may split the integral as the sum of an integral over $(-\frac{\pi}{2}, \frac{\pi}{2})$ and another one over $(\frac{\pi}{2}, \frac{3\pi}{2})$; thus, by dominated convergence it is seen that

$$\int_0^{2\pi} g(r\cos t + \xi(t))e^{it}\, dt \to \int_{-\pi/2}^{\pi/2} g(+\infty)e^{it}\, dt + \int_{\pi/2}^{3\pi/2} g(-\infty)e^{it}\, dt$$

$$= -2[g(+\infty) - g(-\infty)]$$

uniformly on θ. Hence

$$F(r,\theta) \to A_1 + iB_1 + 2e^{i\theta}[(g(+\infty) - g(-\infty)]$$

uniformly on θ. Now regard F as a function of $(x,y) = (r\cos\theta, r\sin\theta)$. If, for example, $g(+\infty) > g(-\infty)$, then the hypothesis implies that the field is outwardly pointing on $\partial B_r(0,0)$ when r is large, and hence F vanishes in $B_r(0,0)$. The general case follows by an approximation argument. Indeed, it suffices to observe that if u is a solution, then the preceding computations give a bound r for $\|u\|_\infty$ depending only on $\|g\|_\infty$ and $\|p\|_\infty$. Next, take $R > r$ large enough and a sequence of smooth functions $g_n : \mathbb{R} \to \mathbb{R}$ satisfying the hypothesis and such that $g_n \to g$ uniformly on $[-R,R]$. The result is now deduced from the Arzelá–Ascoli theorem.

Problems from Chap. 2

2.1 Take, for example, $f : [0, +\infty) \to [0, +\infty)$ given by $f(x) = x + \frac{1}{x+1}$. If X is compact and $d(Tx, Ty) < d(x,y)$ for all $x \ne y$, then define $A = \inf_{x \in X} d(Tx, x)$ and prove that $A = 0$ and that the infimum is achieved.

2.2 Adapt Example 2.1. Observe that f depends also on u', so one should work, for example, on $X = C^1([0,1], \mathbb{R}^n)$.

2.3 Find $\delta > 0$ such that an appropriate fixed point mapping $T : C([-\tau, \delta], \mathbb{R}^n) \to C([-\tau, \delta], \mathbb{R}^n)$ is well defined and contractive. For the second part, use the hint and the contraction mapping theorem, identifying $C([T - \tau, T], \mathbb{R}^n)$ with $C([-\tau, 0], \mathbb{R}^n)$.

2.4 Express the iteration $x_{n+1} = Tx_n$ using integrals.

2.5 Following the hint, observe that $x_n : \mathbb{R} \to \mathbb{R}^n$ is well defined since f_n is bounded. Write the integral expression for x_{n_j} and take limit at both sides to deduce that x is a solution of the problem.

2.6 Observe, in the first place, that if $|x_n(t) - x_0| \leq R$ for all t, then $|x_{n+1}(t) - x_0| \leq M\delta \leq R$. This proves that the sequence is well defined. On the other hand,

$$|x_1(t) - x_0| \leq \int_{t_0}^t |f(s, x_0)| \, ds \leq M(t - t_0).$$

Inductively, call $u_0(t) := M(t - t_0)$, and assume there exists a function u_{n-1} such that $|x_n(t) - x_{n-1}(t)| \leq u_{n-1}(t)$; then

$$|x_{n+1}(t) - x_n(t)| \leq \int_{t_0}^t g(s, |x_n(s) - x_{n-1}(s)|) \, ds \leq \int_{t_0}^t g(s, u_{n-1}(s)) \, ds.$$

Thus, if we define $u_n(t) := \int_{t_0}^t g(s, u_{n-1}(s)) \, ds$, then $|x_{n+1}(t) - x_n(t)| \leq u_n(t)$. Next, observe that

$$u_1(t) = \int_{t_0}^t g(s, u_0(s)) \, ds \leq M(t - t_0) = u_0(t),$$

so we deduce, inductively,

$$u_0(t) \geq u_1(t) \geq u_2(t) \geq \ldots \geq 0.$$

In particular, u_n converges pointwise (and hence uniformly) to some function u satisfying

$$u(t) = \int_{t_0}^t g(s, u(s)) \, ds,$$

that is,

$$u'(t) = g(t, u(t)), \qquad u(t_0) = 0.$$

From the assumptions, $u \equiv 0$. Moreover, from the equality $x'_{n+1}(t) = f(t, x_n(t))$ we deduce that $\|x'_n\|_\infty$ is bounded and, from the Arzelá–Ascoli theorem, there exists a subsequence x_{n_j} converging uniformly to some function x. Since $|x_{n+1} - x_n| \to 0$ uniformly, it follows that x_{n_j+1} also converges uniformly to x, and hence x is a solution of the problem. Finally, observe that if x_n has a subsequence that does not converge uniformly to x, then, using the Arzelá–Ascoli theorem again, it may be assumed that it converges uniformly to some $y \neq x$. This function is also a solution, and $y(t) \in \overline{B_R(x_0)}$ for all t. But in this case $v(t) := |x(t) - y(t)|$ satisfies

$$v(t) \leq \int_0^t g(s, v(s)) \, ds \leq u_0(t)$$

and inductively

$$v(t) \leq \int_0^t g(s, u_{n-1}(s)) \, ds = u_n(t).$$

This proves that $v \equiv 0$, a contradiction.

2.7 If $t_0 := \sup\{t \in [0,1] : \varphi(t) \leq 0\} < 1$, then find $\delta > 0$ such that

$$\varphi(t) \leq \|f(tx + (1-t)y) - f(t_0x + (1-t_0)y)\| - (M+\varepsilon)(t-t_0)\|x-y\| < 0$$

for $t_0 < t < t_0 + \delta$. The second part is easy.

2.8 In the first place, observe that $W(t)$ is continuous, so u_w is well defined. Due to periodicity, $\int_0^1 u'_w(t)\,dt = 0$, so T is also well defined. To prove that $\langle Tw - Tz, w - z \rangle \leq 0$, simply write $w - z = (W - Z)'$ and integrate by parts. The second part follows directly from Theorem 2.4. For the third part, define $F(\lambda, u) := u'' - \lambda f(\cdot, u)$ in appropriate spaces, and prove that $\frac{\partial F}{\partial u}$ is invertible for all u and all $\lambda \geq 0$. If $\lambda_n \in \mathcal{A}$ converge to some λ, take a (unique) sequence of the respective solutions u_n and find the bounds for $\|u'_n\|_{L^2}$ and $\|u'_n\|_\infty$. By the Arzelá–Ascoli theorem, deduce that u_n has a subsequence that converges to a solution for λ.

2.9 Write

$$f(x,y) - f(x_0,y_0) = f(x,y) - f(x_0,y) + f(x_0,y) - f(x_0,y_0)$$

$$= \frac{\partial f}{\partial x}(x_0,y_0)(x - x_0) + R_1(x,y) + \frac{\partial f}{\partial y}(x_0,y_0)(y - y_0) + R_2(x_0,y).$$

Just be a bit careful with R_1... To make things easy, use, for example, the Hahn–Banach and Lagrange theorems.

2.10 For the first part, consider $G : Y \times U \to Y$ given by $G(y,x) := x - F(y)$. For the second part, given f as in Theorem 2.5, define $F : U \times V \to U \times Z$ by $F(x,y) = (x, f(x,y))$, and find $y = y(x)$ from the equality $F(x,y) = (x,0)$.

2.11 There are several ways of proving the result; the most "geometrical" can be sketched as follows. Let $S := \{x \in U : G(x) = 0\}$, and let $c : (-\delta, \delta) \to S$ be a C^1 curve such that $c(0) = x_0$. Because $F \circ c \equiv 0$, it is seen that $DF(x_0)(c'(0)) = 0$. Using the implicit function theorem, this proves that $DF(x_0)$ vanishes over the "tangent space". $T_{x_0}S := \{y \in X : DG(x_0)(y) = 0\} = Ker(DG(x_0))$, and the result follows. Alternatively, one may use the implicit function theorem directly to write the elements of S near x_0 in the form $v + s(z)z$ where $v \in Ker(DG(x_0))$ and $s : Ker(DG(x_0)) \to \mathbb{R}$ is a C^1 function. Finally, it is also possible to reduce the problem to the standard finite-dimensional case.

2.12 The first part is a particular case of (2.3). When $\lambda \leq 0$, the function $-\lambda e^u$ is nonincreasing, so the result was already solved. Positivity follows from the fact that solutions are concave, and for the same reason it is seen that u has a unique absolute maximum at some t_0. Integration over $[0,t_0]$ and $[t_0,1]$ shows that $t_0 = \frac{1}{2}$, and (2.8) is obtained. For the converse, define $u(t)$ with $t \leq \frac{1}{2}$ implicitly as

$$\int_0^{u(t)} \frac{dv}{\sqrt{e^M - e^v}} = \frac{\sqrt{2\lambda t}}{2}$$

and extend it by symmetry to $[0,1]$. Finally, the conclusion follows easily from a study of the left-hand side of (2.8).

2.13 Consider, for example, $X = \{u \in C^2([0,2\pi]) : u(0) = u(2\pi), u'(0) = u'(2\pi)\}$ and $Y = C([0,1])$. Because $F(0,0) = 0$, it suffices to prove that $\frac{\partial F}{\partial u}(0,0)$ is an isomorphism. Verify that $\frac{\partial F}{\partial u}(0,0)\varphi = \varphi'' + a\varphi$ is injective and (consequently) surjective with continuous inverse.

Problems from Chap. 3

3.1 Apply Theorem 3.3.

3.2 Imitate the procedure in Theorem 3.2. Observe that an analogue of Lemma B.3 is needed for the Neumann and periodic conditions.

3.3 Try it with appropriate constants α and β.

3.4 What are the best possible constant values for α and β?

3.5 If $u_{n+1} \neq u_n$ for all n, then prove inductively that $d(u_{n+1}, u_n) = d(Tu_n, Tu_{n-1}) \leq G(d(u_n, u_{n-1})) < d(u_n, u_{n-1})$. Conclude that $\{d(u_{n+1}, u_n)\}$ converges to 0. Moreover, extract a convergent subsequence Tu_{n_j}, and observe that u_{n_j} converges to the same limit. Observe that if $u \neq v$ are fixed points such that $d(u,v) < R$, then $d(u,v) = d(Tu, Tv) < d(u,v)$, a contradiction.
For the second part, consider $Tx(t) := x_0 + \int_{t_0}^t f(s, x(s))\, ds$; then

$$|Tx(t) - Ty(t)| \leq \int_{t_0}^t g(s, |x(s) - y(s)|)\, ds \leq \int_{t_0}^{t_0+\delta} g(s, \|x - y\|_\infty)\, ds := G(\|x - y\|_\infty),$$

with δ as in Exercise 2.6. This defines $G : [0, 2R] \to [0, R]$. If $G(r) \neq r$ for $r \neq 0$, then part 1 applies.

3.6 Apply the diagonal method from Sect. 3.1.4. Moreover, if u and v are two solutions, then $w := u - v$ satisfies $w''w = (u^3 - v^3)(u - v) + w^2 \geq w^2$. Thus, $w'(t)w(t) \geq \int_0^t (w(s)^2 + w'(s)^2)\, dt$, and uniqueness follows.

3.7 Assume, for example, that $u_0 \geq 0$; then any constant $\alpha \ll 0$ is a lower solution of the equation. An upper solution is obtained by considering $\beta \gg 0$ such that $\beta'' \leq 0$, $\beta'(0) \geq u_0$. Deduce the existence of a function u such that $u''(t) = f(t, u(t))$, with $u'(0) = u_0$ and $\alpha \leq u \leq \beta$ on $[0, +\infty)$. Deduce that $u'(u_n) \to 0$ for some sequence $t_n \to +\infty$ and, writing $u'(t) = u'(t_n) + \int_{t_n}^t f(s, u(s))\, ds$, conclude that $u'(t) \to 0$ as $t \to +\infty$.

3.8 Follow the ideas in Sects. 3.2 and 3.3. Note that the conditions for the Newton method are restrictive, but, as seen, for the Dirichlet problem solutions can be obtained using the Newton-continuation method. Also, observe that the quasilinearization method can be applied only in very specific situations. Find an example!

3.9 Make your life easier assuming, without loss of generality, that $u_0 = 0$, $f'(0) = 1$, and $f(0) > 0$. Use the inequality $a - f'(u) \leq Mu$ to deduce that f does not vanish in $[0, 2f(0)]$. A similar analysis yields the existence of one zero in $[-2f(0), 0]$.

3.10 The first part is quite standard and follows applying very carefully the definition of differentiability. Using Hahn–Banach, the result may be reduced to the case $Y = \mathbb{R}$ (in all cases, it may be assumed that $X = \mathbb{R}^2$). Observe, however, that we are not assuming that F of class is C^2. The second part follows directly using the Hahn–Banach Theorem or defining a function

$$\varphi(t) := \|F(ty + (1-t)x) - F(x) - DF(x)(y-x)\| - \frac{M+\varepsilon}{2}t^2\|y-x\|^2$$

and verifying that $\varphi(t) \leq 0$ for $t \in [0,1]$. For the third part, compute the limit $\lim_{s \to 0} \frac{DF(u+sv)(w) - DF(u)(w)}{s}$ with the help of Lagrange's theorem.

3.11 The monotonicity condition ensures that $DF(u)$ is always invertible. Also, observe that solutions are a priori bounded, so f may be replaced by a smooth bounded function.

3.12 Let $v = u'' - K^- u$; then $v'' - K^+ v = u'' - (K^+ + K^-) + K^+ K^- u \geq 0$. This implies $v \leq 0$ and, hence, $u \geq 0$. Using this maximum principle, a quasilinearization method is easily developed. In particular, the equation in the second part is solved taking appropriate constants α and β.

3.13 Direct computation shows that

$$G_\lambda(t,s) = \begin{cases} \frac{1}{2\sqrt{\lambda}}\left(\sigma\cos[\sqrt{\lambda}(t-s)] + \sin[\sqrt{\lambda}(t-s)]\right) & \text{if } t \geq s, \\ \frac{1}{2\sqrt{\lambda}}\left(\sigma\cos[\sqrt{\lambda}(t-s)] - \sin[\sqrt{\lambda}(t-s)]\right) & \text{if } t < s, \end{cases}$$

with

$$\sigma = \frac{\sin(\sqrt{\lambda})}{1 - \cos(\sqrt{\lambda})}.$$

When $\lambda < \frac{\pi^2}{4}$, it is seen that $G_\lambda \geq 0$; thus, if u is 1-periodic and $u'' + \lambda u = \varphi \geq 0$, then

$$u(t) = \int_0^1 G(t,s)\varphi(s)\,ds \geq 0$$

for all t.

In particular, the method of upper and lower solutions can be imitated, now using the operator $u'' + \lambda u$. For the pendulum equation, we may consider the "reverse-order" couple given by $\alpha = \frac{\pi}{2}$ and $\beta = -\frac{\pi}{2}$, and the existence of a solution follows since $\sin' u = \cos u > -\frac{\pi^2}{4}$ for all u. This solution is clearly different from the one obtained earlier using $\beta = \frac{3\pi}{2}$.

Problems from Chap. 4

4.1 Problems 1.4 and 1.5 are essentially the same when generalized to $\mathbb{R}^n$. For problems 1.7–1.9, use Brouwer's theorem in $\mathbb{R}^n$ or, alternatively, Schauder's theorem. Also, one may try to generalize the last results for periodic conditions.

4.2 For $x, y \in \mathbb{R}^n$ let $u_{x,y}$ be the unique solution of the initial value problem

$$u''(t) + g(u(t)) = p(t), \qquad u(0) = x, u'(0) = y,$$

and define

$$F(x,y) = (u'_{x,y}(1) - y, x - u_{x,y}(1)).$$

Using the Poincaré–Miranda theorem, prove that F has a zero.

4.3 If A does not vanish on K, then f is continuous and $f(K) \subset K$, so it has a fixed point.

4.4 For each $n \geq 2$ the second-order difference equation

$$\begin{cases} \Delta^2 y(j) = f(\frac{j}{n}, y(j)), & j = 0, \ldots, n-2 \\ y(0) = y(n) = 0 \end{cases}$$

has at least one solution y_n. Define $z_n : \{\frac{j}{n} : j = 0, \ldots, n\} \to \mathbb{R}$ by $z_n(t) = y_n(nt)$, extend it to a piecewise linear function on $[0,1]$, and prove the existence of a subsequence that converges uniformly.

4.5 The first two items follow by direct computation. Note that r is well defined since $h_{\lambda(f)}(f)$ does not vanish. The obvious choice for a mapping without fixed points is $Tx := -r(x)$.

4.6 If $\|x\| \leq 1$, then $\|f(x)\| \leq \frac{1-\|x\|}{2} + \|x\| \leq 1$. Moreover, if x is a fixed point of f, then $x_n = x_{n-1}$ for all $n \neq 0$. Since $\sum_{n \in \mathbb{Z}} x_n^2$ converges, it follows that $x = 0$. On the other hand, $x_0 = \frac{1-\|x\|}{2} + x_{-1}$, a contradiction.

4.7 Prove that if $x(0) + x(1) = 0$, then $\|x\|_\infty \leq \frac{1}{2} \int_0^1 |x'(t)| \, dt$, and use Schauder's theorem.

4.8 Use Schauder's theorem in $C^1([0,1])$.

4.9 For the first part, just imitate the procedure in Sect. 4.3. For the second part, start, for example, with a simple case: if α and β are T-periodic vector functions such that $\alpha_i \leq \beta_i$ for $i = 1, \ldots, n$ and

$$\alpha_i''(t) \geq f_i(t, u_1, \ldots, \alpha_i(t), \ldots, u_n), \qquad \beta_i''(t) \leq f_i(t, u_1, \ldots, \beta_i(t), \ldots, u_n)$$

for all t, all i, and $\alpha_j(t) \leq u_j \leq \beta_j(t)$, then the problem $u''(t) = f(t, u(t))$ has a T-periodic solution u with $\alpha_i \leq u_i \leq \beta_i$ for all i. If the equation also depends on u', then try to find an accurate Nagumo-type condition.

4.10 For the first part, just use Lagrange's theorem. The second part is proven, as usual, multiplying by $v - \bar{v}$ and integrating. The third part follows from direct computation and yields the explicit condition $M(P_0) > M$.

4.11 The proof is similar to that in Sect. 4.3.1, although in this case, one obtains

$$\int_r^R \frac{ds}{\psi(s)} \leq \left| \int_{t_0}^{t_1} ds \right| = |t_1 - t_0| \leq 1.$$

4.12 Following the hint, if, for example, $u'(t) > 0$ in (t_0, t_1), then consider the inverse function $t : (u(t_0), u(t_1)) \to (t_0, t_1)$. Using the fact that $q'(u) = 2p(u)p'(u) = 2u''(t) = f(t, u(t), u'(t))$, deduce that $q'(u) \leq 2c(u)(1 - q(u))$, and hence $q(u) \leq e^C - 1$ for some constant C depending only on k. If u' does not vanish in $[0, 1]$, then consider, as before, $t = t(u)$. From the inequality $q'(u) \leq 2c(u)(1 + q(u))$ it is seen for all ξ that

$$\left| \log \frac{1 + q(u)}{1 + q(u(\xi))} \right| \leq C$$

for some C depending only on k. Using Lagrange's theorem, $u(1) - u(0) = u'(\xi) = p(u(\xi))$ for some ξ, and hence $|q(u(\xi))| \leq \sqrt{2k}$.

4.13 Nonexistence for large T follows by simple computation. There is no contradiction since nobody said that a couple $\alpha \leq \beta$ of a lower and an upper solution exists. Observe that $\alpha \equiv 0$ is a lower solution, which implies there is no upper solution $\beta \geq 0$.

4.14 The first part follows multiplying by $u - v$. For the particular case, just take v as the segment joining the points $(0, u(0))$ and $(1, u(1))$. Concerning the second part, existence and uniqueness are proven in previous examples. The continuity of $\mathcal{T}_0$ is deduced from the bounds obtained in 1. For the third part, for each $v \in C([0, 1])$ define $p_v(t) := g(t, v(t))$; then consider the mapping $Kv := \mathcal{T}_0(p_v) \in C^2([0, 1]) \hookrightarrow C([0, 1])$ and use Schauder's theorem. Part 4 generalizes 2 and follows essentially in the same way; however, since the boundary conditions are now variable, the bounds in 1 will include an extra term. Finally, 5 follows since, for p fixed, $\mathcal{T}(p, \cdot) : \mathbb{R}^2 \to C^2([0, 1])$ is injective and continuous. The inverse is the trace function $u \mapsto (u(0), u(1))$, which is obviously continuous.

Problems from Chap. 5

5.1 Take $y \in \partial B_1(0) \backslash f(\partial B_1(0))$, and consider the homotopy $h(x, \lambda) = \lambda f(x) - y$.

5.2 Prove the statement firstly assuming that f is smooth and 0 is a regular value.

5.3 Write $f(z) = \sum a_n z^n$ with $a_n = \frac{f^{(n)}(0)}{n!}$, and compute $\frac{1}{2\pi} \int_{\partial B} \frac{f'(z)}{f(z)} dz$.

5.4 If f has no fixed points on the boundary, then there exists a homotopy between $f(x) - x$ and the function $x - x_0$.

5.5 Over large balls, $z^n - P(z, \bar{z})$ is homotopic to z^n.

5.6 The extension of φ to a continuous odd function with compact support can be proven by induction and Tietze's theorem. The uniform convergence $\varphi * \phi_\varepsilon \to \varphi$ follows from the standard theory of mollifiers. For item 2 it suffices to define $g(x) = Ax + R(x)$ with $A \in R^{n \times n}$ invertible such that $|Ax - D\tilde{f}(0)x| < \varepsilon - \|f - \tilde{f}\|_\infty$ for all x. The proof of the fact that invertible matrices are dense in $\mathbb{R}^{n \times n}$ is an easy exercise.

5.7 Use the homotopies $h_\pm(x, \lambda) = \lambda f(x) \pm (1 - \lambda)x$.

5.8 Let x be the unique solution of the problem with initial value x_0. Verify that $x(t) = x_0 e^{-t} + \int_0^t e^{s-t} G(s, x(s)) \, ds$, and, in particular, if $|x_0|$ is large, then $\|x\|_\infty \le 2|x_0|$. Given an appropriate $\varepsilon > 0$, enlarge $|x_0|$ if necessary to deduce that $|P(x_0) - \frac{x_0}{e}| < \varepsilon|x_0|$ and conclude that $I \pm P$ is homotopic to the identity.

5.9 Define an appropriate truncation function, and verify that $I + P$ has a zero, where P is the Poincaré operator. The conclusion can also be deduced from Schauder's theorem or directly by Schaefer's theorem, as proposed in Exercise 4.7 for a sublinear problem.

5.10 Multiplying the equation $u'(t) + \nabla F(u(t)) = 0$ by $u'(t)$, it is seen that $F(u(t)) \le F(u_0)$ for all t. Furthermore, if $F(u(t)) = F(u_0)$ for some $t > 0$, then $u \equiv u_0$ in a neighborhood of 0 and u_0 is a critical point of F. This implies that if $|u_0| = r \ge R$, then $F(u(t)) < F(u_0)$ for all $t > 0$. In particular, u is defined for all $t > 0$ and $u(t) \ne u_0$. Thus, the Poincaré operator P_λ does not have fixed points on $\partial B_r(0)$ and, as in (5.11), it is seen that $deg(\nabla F, B_r(0), 0) = deg(Id - P_\lambda, B_r(0), 0)$. Define $M := \max_{|u| \le R} F(u)$, and fix $r > R$ such that $F(u) > M$ for $|u| \ge r$. Next, define $\tilde{M} := \max_{|u| = r} F(u)$, and fix $\tilde{r}$ such that $F(u) > \tilde{M}$ for $|u| \ge \tilde{r}$. Let $|u_0| = r$, and assume $|u(t)| \ge R$ for all $t \le \lambda$. Because

$$F(u(t)) = F(u_0) - \int_0^t |\nabla F(u(s))|^2 \, ds,$$

for all $t \le \lambda$ it is deduced, in the first place, that $|u(t)| < r$ and, in the second place, that

$$F(u(t)) \le \tilde{M} - \gamma t,$$

where $\gamma := \inf_{R \le |u| \le \tilde{r}} |\nabla F(u)|$. Taking $\lambda = \frac{\tilde{M} - M}{\gamma}$ we draw the following conclusions:

- If $|u(t)| \ge R$ for all $t \le \lambda$, then $F(u(\lambda)) \le M$ and, hence, $|u(\lambda)| < r$.
- If $|u(t)| < R$ for some $t \le \lambda$, then $F(u(\lambda)) \le F(u(t)) \le M$ and, again, $|u(\lambda)| < r$.

In particular, $Id - \sigma P_\lambda$ does not vanish on $\partial B_r(0)$ for $0 \le \sigma \le 1$, which implies $deg(Id - P_\lambda, B_r(0), 0) = 1$.

5.11 The function $h(v_0, \lambda) = u_{\lambda, v_0}(1)$ is well defined and continuous and does not vanish for $v_0 \in \partial\Omega$. Compute $h(v_0, 0)$.

5.12 Use the fact that the Hilbert cube is compact and the Poincaré–Miranda theorem.

5.13 Given a continuous function $f : \overline{B} \to \overline{B}$ with no fixed points, it is easy to find a retraction. For the second item, assume that B is centered at 0 and consider, for example, $r(\lambda x)$.

5.14 Given $y \in E$, prove that $deg(I - K, B_R(0), y) \neq 0$ for R large enough.

5.15 If K is compact and odd, and $\tilde{K}_\varepsilon$ is an ε-approximation of K, then $K_\varepsilon(x) := \frac{\tilde{K}_\varepsilon(x) - \tilde{K}_\varepsilon(-x)}{2}$ is an odd ε-approximation of K. All the consequences of Borsuk's theorem are obtained as in the finite-dimensional case.

5.16 Since $I - K$ is odd, its degree over any ball centered in 0 is nonzero. In particular, its image contains a neighborhood of 0 and, by homogeneity, $Im(I - K) = X$. Continuity of $(I - K)^{-1}$ can be proven as follows. Suppose that $y_n = (I - K)(x_n) \to 0$ and $\|x_n\| \to \infty$, and consider $z_n = \frac{x_n}{\|x_n\|}$. Then $(I - K)(z_n) \to 0$, and, by compactness, we may assume that $Kz_n \to z$ for some z. It follows that $z_n \to z \neq 0$ and $Kz = z$, a contradiction since 0 is the unique fixed point of K.

5.17 Apply a generalized version of Rouché's theorem.

5.18 Use the extension of Theorem 5.6 obtained in Exercise 5.15: let $E = \{u \in C([0,1]) : u(0) = u(1) = 0\}$, and define $F(p,u) = u - K(p,u)$, where $K(p,u)$ is the unique solution $v \in E$ of the problem $v''(t) = p(t) - g(t,u(t))$. From the fact that $F(0,0) = 0$ and the hypothesis, the existence of a unique $u = u(p)$ is deduced for $\|p\|_\infty$ small enough.
Alternative proof: define $K : E \to E$ by $K\varphi = u$, where u is the unique solution of the linear problem $u'' = \varphi$. The equation $u = K(p - g(\cdot, u))$ in E is equivalent to $F(u) = Kp$, where $F(u) := u + K(g(\cdot, u))$. Because F is injective in $X = B_r(0)$ and $F(0) = 0$, it is deduced that $F(X)$ contains a neighborhood of 0.

5.19 The first part follows exactly as in Exercise 5.4. For the second part, prove that, in all cases, if $|x| = 1$ and $t > 1$, then $Kx \neq tx$.

5.20 For the first part, define the operator $Kx(t) := x_0 + \int_{t_0}^t f(s, x(s)) ds$ in a convenient space, and compute its degree over a neighborhood of x_0. For the second part, define Kv as the unique solution of the linear problem $u''(t) = f(t, v(t))$ under Dirichlet conditions.

5.21 Let $K : C([0,1]) \to C([0,1])$ be the compact operator given by $Kv = u$, where u is the unique solution of the problem

$$u''(t) - u(t) = p(t) - g(v(t)) - v(t), \qquad u(0) = u(1), u'(0) = u'(1).$$

If $u = \lambda Ku$ for some $\lambda \in (0,1)$, then

$$-u''(t) + (1 - \lambda)u(t) = \lambda[g(u(t)) - p(t)].$$

Multiply by u and integrate to obtain

$$\int_0^1 u'(t)^2 \, dt + (1-\lambda) \int_0^1 u(t)^2 \, dt = \lambda \int_0^1 [g(u(t)) - p(t)] u(t) \, dt.$$

Because $\bar{p} = 0$, deduce that $\int_0^1 p(t) u(t) \, dt = \int_0^1 p(t) [u(t) - \bar{u}] \, dt$ and, hence,

$$\int_0^1 u'(t)^2 \, dt + (1-\lambda) \int_0^1 u(t)^2 \, dt \leq C + \|p\|_{L^2} \|u - \bar{u}\|_{L^2}$$

for some constant C. Deduce that $\|u - \bar{u}\|_\infty$ is bounded, and moreover, taking average in the equation it follows that

$$(1-\lambda)\bar{u} = \lambda \int_0^1 g(u(t)) \, dt.$$

Thus, $\bar{u} \int_0^1 g(u(t)) \, dt \geq 0$, and writing $g(u(t)) = g(\bar{u} + u(t) - \bar{u})$ conclude that $\bar{u}$ cannot be arbitrarily large.

A counterexample when the inequalities are reversed can be constructed imitating the one given in Remark 1.1 for Dirichlet conditions. For an equation with the friction term $c(t)u'(t)$, bounds for u' are also obtained as in Chap. 1 (Remark 1.6): multiply both sides by $u'(t)$ and integrate to obtain

$$\int_0^1 c(t) u'(t)^2 \, dt = \lambda \int_0^1 p(t) u'(t) \, dt,$$

and hence

$$\|u'\|_{L^2} \leq \frac{1}{\inf_{t \in [0,1]} |c(t)|} \|p\|_{L^2}.$$

Problems from Chap. 6

6.1 In all cases, a priori bounds can be found for the problem

$$u''(t) = \lambda f(t, u(t)), \qquad u(0) = u_0, \ u(1) = u_1,$$

so degree theory can be applied. If f depends also on u', then try to find a Nagumo-type condition.

6.2 The first item follows directly from the monotonicity of g. For the second item, consider as in Sect. 6.1.2 the spaces $C_T = \{u \in C(\mathbb{R}) : u(t+T) = u(t) \text{ for all } t\}$ and $\tilde{C}_T = \{u \in C_T : \bar{u} = 0\}$. Given $\varphi \in \tilde{C}_T$, there exists a unique $u \in \tilde{C}_T$ such that $u'(t) = \varphi(t)$ and the operator $K : \varphi \mapsto u$ is compact. Let $N : C_T \to C_T$ be given by $Nu(t) = -a(t) + g(t, u(t-\tau))$. For $\lambda > 0$ the periodic problem for the equation $u' = \lambda Nu$ is equivalent to the fixed point equation $u = \bar{u} + \overline{Nu} + \lambda K(Nu - \overline{Nu})$. Thus, it suffices to prove that, for some $R > 0$,

1. $u' \neq \lambda N u$ for all $\lambda \in (0,1)$ and $u \in C_T$ such that $\|u\|_\infty = R$,
2. $[\int_0^T a(t)\,dt - \phi(-R)][\int_0^T a(t)\,dt - \phi(R)] < 0$.

From the monotonicity of ϕ, the latter condition is trivial if $\int_0^T a(t)\,dt \in Im(\phi)$. For the first one, suppose that $u \in C_T$ is such that $u'(t) = \lambda(-a(t) + g(t, u(t-\tau)))$; then integrate at both sides and, because g is nonincreasing in u, we obtain

$$\int_0^T a(t)\,dt = \int_0^T g(t, u(t-\tau))\,dt \geq \int_0^T g(t, u_{max})\,dt = \phi(u_{max}).$$

A similar computation yields $\int_0^T a(t)\,dt \leq \phi(u_{min})$, and hence

$$u_{min} \leq \phi^{-1}\left(\int_0^T a(t)\,dt\right) \leq u_{max}.$$

In particular, there exists t_0 such that $u(t_0) = \phi^{-1}\left(\int_0^T a(t)\,dt\right)$. On the other hand, using the fact that a and g are positive, deduce that $|u'(t)| < \lambda(a(t) + g(t, u(t-\tau)))$ and

$$\int_0^T |u'(t)|\,dt < \int_0^T a(t) + g(t, u(t-\tau))\,dt = 2\int_0^T a(t)\,dt.$$

This proves that $|u(t) - u(t_0)| = \left|\int_{t_0}^t u'(t)\,dt\right| \leq 2\int_0^T a(t)\,dt$, and a bound for $u(t)$ is obtained.

6.3 For the upper bounds, use the fact that $xe^{-x} \leq \frac{1}{e}$. For the lower bounds, use the fact that xe^{-x} is nondecreasing for $x < 1$.

6.4 Try any other example with a one-dimensional kernel, for example, a second-order equation with Robin conditions

$$u'(0) = a_0 u(0), \qquad u'(1) = a_1 u(1)$$

for some appropriate a_0 and a_1. Also, the nth-order equation

$$u''(t) = f(t, u(t), \ldots, u^{n-1}(t))$$

with periodic conditions

$$u(0) = u(1), \ldots, u^{n-1}(0) = u^{n-1}(1)$$

can be solved under a Landesman–Lazer-type condition.

6.5 The first part follows as a consequence of the general setting in Theorem 6.4. If u_k is periodic such that $u_k'' = \lambda_k(p - g(u_k))$ for $\lambda_k \in (0,1)$ and $\|u_k\| \to \infty$, then $\|u_k - \overline{u_k}\|_\infty$ is bounded and $\overline{u_k} \to \infty$. Taking a subsequence, it may be assumed that $\frac{\overline{u_k}}{|\overline{u_k}|}$ converges to some $v \in S^{n-1}$, and using Fatou's lemma from the identity $\overline{p} =$

$\int_0^1 g(u_k(t)) \, dt$ a contradiction is obtained. Observe that Nirenberg's theorem is a particular case of this result: for each $v \in S^{n-1}$ take a neighborhood $U_v \subset S^{n-1}$ such that $\langle g_u - \overline{p}, g_v - \overline{p} \rangle > 0$ and $w_v := -\frac{g_v - \overline{p}}{|g_v - \overline{p}|}$. The family $\{U_v\}$ covers the sphere, so the first condition holds. The second condition is obviously the same. Finally, let $r = \frac{\|p\|_\infty + \|g\|_\infty}{4\pi}$, and take R large enough such that for all $v \in \partial B_R(0)$ there exists j with $\overline{B_r(v)} \subset C(U_j)$, where $C(U_j) := \{su : u \in U_j, s > 0\}$. Enlarging R if necessary, we may suppose that $\langle g(y) - \overline{p}, w_j \rangle < 0$ for all $y \in \overline{B_r(v)}$, which implies that $\overline{p} \notin co(g(\overline{B_r(v)}))$.

6.6 The general setting is very similar to the Lazer–Leach case, although bounds must be obtained for

$$u''(t) + u(t) = \lambda \left(p(t) - [F(u(t))]' \right)$$

with $0 < \lambda < 1$. The original proof in [37] employed a shooting-type argument (and Brouwer's theorem in the plane). For a short sketch of the proof see also [69]. The problem is essentially different from that in Example 4.5, which is nonresonant since $T < 2\pi$.

6.7 Set

$$d_j := \inf_{u \in \mathbb{R}^n} g(u) \cdot w_j, \qquad v_j := d_j w_j,$$

where $\{w_1, \ldots, w_n\} \subset S^{n-1}$ is a basis of $\mathbb{R}^n$ such that $(g(u) - \xi) \cdot w_j \geq 0$ for every $u \in \mathbb{R}^n$. Then

$$(g(u) - v_j) \cdot w_j \geq d_j - v_j \cdot w_j = 0.$$

Thus,

$$\left| u''(t) \cdot w_j \right| \leq (g(u) - v_j) \cdot w_j + \left| (v_j - p) \cdot w_j \right|$$

and, consequently,

$$\int_0^1 \left| u''(t) \cdot w_j \right| \, dt \leq \int_0^1 \left| (v_j - p) \cdot w_j \right| \, dt - \xi_j \cdot w_j := K_j.$$

Hence, for each $t \in [0, 1]$ we deduce

$$\left| u'(t) \cdot w_j \right| \leq K_j,$$

and the desired bounds are obtained.

6.8 Write

$$u'(t) = u(1) - u(0) + \int_0^t su''(s) \, ds - \int_t^1 (1 - s)u''(s) \, ds;$$

then using the hypothesis

$$|u'(t)| \leq |u(1) - u(0)| + \int_0^t s\left[\left(c(u(s) \cdot f(s, u(s), u'(s)) + |u'(s)|^2\right) + K\right] ds$$

$$+ \int_t^1 (1-s)\left[\left(c(u(s) \cdot f(s, u(s), u'(s)) + |u'(s)|^2\right) + K\right] ds$$

$$\leq A + c\int_0^t s\left(u(s) \cdot u''(s) + |u'(s)|^2\right) dt + c\int_t^1 (1-s)\left(u(s) \cdot u''(s) + |u'(s)|^2\right) dt,$$

where $A = |u(1) - u(0)| + \frac{K}{4} \leq 2R + \frac{K}{4}$. Next, define $\phi(t) := \frac{|u(t)|^2}{2}$; then

$$|u'(t)| \leq A + c\left(\int_0^t s\phi''(s)\, ds + \int_t^1 (1-s)\phi''(s)\, ds\right).$$

Finally, integration by parts yields

$$|u'(t)| \leq A + c(2t-1)\phi'(t) + c(\phi(1) - \phi(0)) = A + c(2t-1)u(t) \cdot u'(t) + \frac{cR^2}{2}.$$

Since $|u(t)| \leq R$, we conclude that

$$|u'(t)| \leq A + \frac{cR^2}{2} + (2t-1)cR|u'(t)| \leq A + \frac{cR^2}{2} + cR|u'(t)|,$$

and the result follows since $cR < 1$. Once a bound for $\|u'\|_\infty$ is thus obtained, the existence result follows from the standard continuation method.

6.9 Let $R = R_1 + \frac{\|p\|_{L^2}}{|a|}$ and $\Omega := \{u \in C([0,1]) : R_0 < \|u\|_\infty < R\}$. For $0 < \lambda < 1$, if

$$u''(t) + au'(t) = \lambda(p(t) - \nabla G_n(u(t))), \qquad u(0) = u(1),\ u'(0) = u'(1),$$

then, as seen in previous cases, $\|u'\|_{L^2} < \frac{\|p\|_{L^2}}{|a|}$, and hence $\|u - \bar{u}\|_\infty < \frac{\|p\|_{L^2}}{2|a|} := r$.

Suppose that $\|u\|_\infty = R_0$, and let $\psi(t) := \frac{|u(t)|^2}{2}$. Take $t_0 \in [0,1)$ such that $\psi(t_0) = \frac{R_0^2}{2}$; then $\psi'(t_0) = 0$ and $\psi''(t_0) \leq 0$. This implies $u'(t_0) \cdot u(t_0) = 0$ and

$$0 \geq u''(t_0) \cdot u(t_0) = u''(t_0) \cdot u(t_0) + au'(t_0) \cdot u(t_0) = \lambda[p(t_0) \cdot u(t_0) - \nabla G_n(u(t_0)) \cdot u(t_0)].$$

From the first assumption, the latter term is strictly positive, a contradiction. Next, suppose that $\|u\|_\infty = R$; then

$$|\bar{u}| \geq \|u\|_\infty - \|u - \bar{u}\|_\infty > R - r = R_1 + r.$$

Moreover, from the periodic conditions

$$0 = \int_0^1 p(t)\, dt = \int_0^1 \nabla G_n(u(t))\, dt$$

and, as $u(t) \in B_r(\overline{u})$ for all t, it is seen that $|u(t)| > R_1$, and hence $G_n(u(t)) \cdot w > 0$ for all t, a contradiction.

Finally, note that the function ϕ defined in Theorem 6.4 coincides with $-\nabla G_n$ and, by excision,

$$deg(\nabla G_n, B_R(0) \backslash B_{R_0}(0), 0) = deg(\nabla G_n, B_R(0), 0) - deg(\nabla G_n, B_{R_0}(0), 0) = (-1)^{n+1}$$

since $\phi(u) \cdot u < 0$ for $|u| = R_0$. The existence of u_n is thus proven. From the fact that $\|u_n'\|_{L^2}$ is bounded, by Arzelá–Ascoli there exists a subsequence converging uniformly to some u, and clearly $\|u\|_\infty \geq R_0$.

It is easily verified that if u does not vanish over an open interval I, then u solves the differential equation over I. Finally, suppose that $u(t_1) = 0$ for some t_1. Because $u \not\equiv 0$, we may suppose, for example, that u does not vanish in $I := [t_0, t_1)$ for some $t_0 < t_1$. Taking t_0 closer to t_1 if necessary, we may also assume that $G(u(t)) > 0$ for $t \in I$. We know that $u'' + au' + \nabla G(u) = p$ over I, and multiplying by u' we deduce

$$\frac{|u'(t)|^2}{2} + G(u(t)) = \frac{|u'(t_0)|^2}{2} + G(u(t_0)) + \int_{t_0}^t p(s)u'(s)\,ds \qquad \text{(B.2)}$$

for all $t \in I$. Thus, for any fixed $\hat{t} < t_1$,

$$\max_{t_0 \leq t \leq \hat{t}} \frac{|u'(t)|^2}{2} < A + B \max_{t_0 \leq t \leq \hat{t}} |u'(t)|$$

for some constants A and B independent of $\hat{t}$. This implies that $\sup_{t_0 \leq t \leq t_1} |u'(t)| \leq M$ for some constant M, and letting $t \to t_1$ in (B.2) yields a contradiction. This proves that u does not vanish, and the proof is complete.

References

[1] R. Agarwal and D. O'Regan: Infinite Interval Problems for Differential, Difference and Integral Equations. Kluwer (2001).

[2] S. Ahmad, A. Lazer, J. Paul, Elementary critical point theory and perturbation of elliptic boundary value problems at resonance. Indiana Univ. Math. J., 25 (1976), 933–944.

[3] H. Amann: Ordinary Differential Equations. An Introdution to Nonlinear Analysis. Walter de Gruyter Berlin. New York (1990).

[4] A. Ambrosetti, A. Malchiodi: Nonlinear Analysis and Semilinear Elliptic Problems. Cambridge Studies in Advanced Mathematics 104, Cambridge University Press (2007).

[5] S. Banach: Sur les opérations dans les ensembles abstraits et leur application aux équations intégrales. Fund. Math. 3 (1922), 133–181.

[6] P. Amster and M. Clapp: Periodic solutions of resonant systems with rapidly rotating nonlinearities. Discrete and Continuous Dynamical Systems, Series A 31 No. 2 (2011), 373–383.

[7] P. Amster and A. Déboli: Existence of positive T-periodic solutions of a generalized Nicholson's blowflies model with a nonlinear harvesting term. Applied Mathematics Letters 25 No. 9 (2012), 1203–1207.

[8] P. Amster and P. De Nápoli: Non-asymptotic Lazer–Leach type conditions for a nonlinear oscillator. Discrete and Continuous Dynamical Systems, Series A 29, No. 3 (2011), 757–767.

[9] P. Amster and J. Haddad: A Hartman-Nagumo type condition for a class of contractible domains. Topological Methods in Nonlinear Analysis 41, No. 2 (2013), 287–304.

[10] P. Amster, L. Idels: Periodic Solutions in General Scalar Nonautonomous Models with Delays. Nonlinear Differential Equations and Applications NoDEA 20 (2013), 1577–1596.

[11] P. Amster, M. K. Kwong and C. Rogers: On a Neumann Boundary Value Problem for Painlevé II in Two Ion Electro-Diffusion. Nonlinear Analysis, TMA 74, 9 (2011), 2897–2907.

[12] P. Amster and M. Maurette: Periodic solutions of systems with singularities of repulsive type. Advanced Nonlinear Studies 11 (2011), 201–220.

[13] P. Bates, Reduction theorems for a class of semilinear equations at resonance. Proceedings of the American Mathematical Society 84, 1, (1982), 73–78.

[14] J. Bebernes and L. Jackson: Infinite interval boundary value problems for $y'' = f(x,y)$. Duke Math. J. 34 (1967), 39–47.

[15] J. Bebernes, K. Schmitt: Periodic boundary value problems for systems of second order differential equations. Journal of Differential Equations 13, no. 1 (1973) 32–47.

[16] R. Bellman and R. Kalaba: Quasilinearisation and Nonlinear Boundary Value Problems. American Elsevier, New York (1965).

[17] P. Bohl: Über die Bewegung eines mechanischen Systems in der Nähe einer Gleichgewichtslage, J. Reine Angew. Math. 127 (1904), 179–276.

[18] K. Borsuk: Sur les rétractes, Fund. Math. 17 (1931), 152–170.

[19] L. Brouwer: Über Abbildungen von Mannigfaltigkeiten, Math. Ann. 71 (1912), 97–115.

[20] L. Brouwer: Beweis der Invarianz der Dimensionenzahl, Math. Ann. 70 (1911), 161–165.

[21] F. Browder: Nonexpansive nonlinear operators in a Banach space, Proc. Natl. Acad. Sci. U.S.A. 54 (1965), 1041–1044.

[22] R. Brown: A Topological Introduction to Nonlinear Analysis. Birkhaser. Boston - Basel - Berlin (1993).

[23] R. Brooks and K. Schmitt: The contraction mapping principle and some applications, Electronic Journal of Differential Equations, Monograph 09 (2009).

[24] L. Carroll: What the Tortoise Said to Achilles. Mind, 4 (1895), 278–80.

[25] A. Castro: Periodic solutions of the forced pendulum equation. Diff. Equations (1980), 149–60.

[26] J. Cronin: Fixed Points and Topological Degree in Nonlinear Analysis. Math. Surveys No.11, Amer. Math. Soc. Providence R.I (1964).

[27] K. Deimling: Nonlinear functional analysis. Springer, Berlin, Heidelberg, New York, Tokyo (1985).

[28] C. De Coster and P: Habets, Two-point Boundary Value Problems: Lower and Upper Solutions, Mathematics in Science and Engineering, 205, Elsevier B. V., Amsterdam, 2006.

[29] C. De Coster, and P. Habets, Upper and Lower Solutions in the Theory of ODE Boundary Value Problems: Classical and Recent Results. In Nonlinear Analysis and Boundary Value Problems for Ordinary Differential Equations, F. Zanolin ed., Springer, 1996, CISM Courses and Lectures, 371.

[30] G. Dinca, J. Mawhin: Brouwer degree and applications. Preprint (2009).

[31] M. do Carmo: Riemannian Geometry, Birkhäuser (1992).

[32] P. Drabek and J. Milota: Methods of Nonlinear Analysis. Applicatons to Differential Equations, Birkhuser (2007).

[33] J. Dugundji: An extension of Tietze's theorem. Pacific J. Math. 1 No. 3 (1951), 353–367.

[34] C. Fabry and C. Franchetti: Nonlinear equations with growth restrictions on the nonlinear term. J. Differential Equations 20 No. 2 (1976), 283–291.

[35] I. Fonseca, W. Gangbo: Degree Theory in Analysis and Applications. Oxford Lecture Series in Mathematics and Its Applications, Clarendon Press, Oxford (1995).

[36] G. Fournier and J. Mawhin: On periodic solutions of forced pendulum-like equations. Journal of Differential Equations 60 (1985) 381–395.

[37] P. Frederickson and A. Lazer: Necessary and sufficient damping in a second order oscillator, J. Differential Eqs. 5 No. 2 (1969), 262–270.

[38] S. Fučík and M. Krbec: Boundary value problems with bounded nonlinearity and general null-space of the linear part. Mathematische Zeitschrift 155, 2 (1977), 129–138.

[39] D. Gale: The game of hex and the Brouwer fixed-point theorem. American Mathematical Monthly 86, No. 10 (1979), 818–827.

[40] A. Granas, R. Guenther, J. Lee and D. O'Regan. Boundary Value Problems on Infinite Intervals and Semiconductor Devices. Journal of Math. Analysis and Applications 116 (1986), 335–348.

[41] R. Gaines and J. Mawhin: Coincidence degree and nonlinear differential equations, Lecture Notes in Mathematics 586, Springer, Berlin (1977).

[42] R. Gaines and J.Mawhin: Ordinary differential equations with nonlinear boundary conditions. J. Differential Equations 26 (1977) 200–222.

[43] P. Habets and R. Pouso: Examples of the nonexistence of a solution in the presence of upper and lower solutions. ANZIAM J. 44 (2003), 591–594.

[44] P. Hartman: Ordinary differential equations. Wiley, New York (1964).

[45] P. Hartman: On boundary value problems for systems of ordinary nonlinear second order differential equations. Trans. Amer. Math. Soc. 96 (1960), 493–509.

[46] S. Hastings and J.B. McLeod: Short proofs of results by Landesman, Lazer, and Leach on problems related to resonance. Differential and integral equations 24, No. 5–6 (2011), 435–441.

[47] S. Hastings and J.B. McLeod: Classical Methods in Ordinary Differential Equations: With Applications to Boundary Value Problems. Graduate Studies in Mathematics 129, American Mathematical Society (2012).

[48] S. Heikkilä and V. Lakshmikantham: Monotone Iterative Techniques for Discontinuous Nonlinear Differential Equations. Marcel Dekker (1994).

[49] E. Heinz: An elementary analytic theory of the degree of mapping in n-dimensional space. J. Math. and Mech. 8 (1959), 231–247.

[50] M. Henle: A Combinatorial Introduction to Topology. W.H. Freeman and Company: San Francisco (1979).

[51] P. Hess: On a theorem by Landesman and Lazer. Indiana Univ. Math. J. 23 No. 9 (1974), 827–829

[52] S. Kakutani: Topological properties of the unit sphere of a Hilbert space, Proc. Imp. Acad. Tokyo 19 (1943), 269–71.

[53] Y. Kannai: An elementary proof of the no-retraction theorem. American Mathematical Monthly Vol. 88, No. 4 (1981), 264–268.

[54] R. Kannan and K. Nagle: Forced oscillations with rapidly vanishing nonlinearities. Proc. Amer. Math. Soc. 111 (1991), 385–393.

[55] R. Kannan and R. Ortega: Periodic solutions of pendulum-type equations. J. Differential Equations 59 (1985), 123–144.

[56] H. Knobloch: On the existence of periodic solutions of second order vector differential equations, J. Differential Equations 9 (1971), 67–85.

[57] S. Krantz, H. Parks: The implicit function theorem: history, theory, and applications. Birkhauser, Boston (2002).

[58] M. A. Krasnoselskii: Fixed points of cone-compressing or cone-extending operators. Soviet Math. Dokl. 1 (1960) 1285–1288.

[59] M. A. Krasnoselkii and A. Perov: On a certain principle for the existence of bounded, periodic and almost periodic solutions of systems of ordinary differential equations (Russian), Dokl. Akad. Nauk SSSR 123 (1958), 235–238.

[60] L. Kronecker, Ueber Systeme von Funktionen mehrer Variablen. Monatsberichte (1869) Erster Abhandlung 159–193, Zweite Abhandlung, 688–698.

[61] W. Kulpa: The Poincaré–Miranda Theorem. American Mathematical Monthly 104, No. 6 (1997), 545–550.

[62] V. Lakshmikantham: Monotone Approximations and Rapid Convergence. Ten Mathematical Essays on Approximation in Analysis and Topology J. Ferrera, J. López-Gómez, F. R. Ruiz del Portal, Editors (2005), 125–149.

[63] V. Lakshmikantham: An extension of the method of quasilinearization. J. Optim. Theory Appl. 82 (1994) 315–321.

[64] V. Lakshmikantham, Further improvement of generalized quasilinearization. Nonlinear Analysis, 27 (1996) 315–321.

[65] V. Lakshmikantham, A. S. Vatsala: Generalized quasilinearization for nonlinear problems. Mathematics and its Applications, 440. Kluwer Academic Publishers, Dordrecht (1998).

[66] V. Lakshmikantham: Monotone Approximations and Rapid Convergence. Ten Mathematical Essays on Approximation in Analysis and Topology J. Ferrera, J. López-Gómez, F. R. Ruiz del Portal, Editors (2005), 125–149.

[67] E. Landesman and A. Lazer: Nonlinear perturbations of linear elliptic boundary value problems at resonance, J. Math. Mech. 19 (1970), 609–623.

[68] A. Lazer: On Schauder's Fixed point theorem and forced second-order nonlinear oscillations, J. Math. Anal. Appl. 21 (1968) 421–425.

[69] A. Lazer: A second look at the first result of Landesman-Lazer type. Electronic Journal of Differential Equations 2000 (2000), 113–119.

[70] A. Lazer and D. Leach: Bounded perturbations of forced harmonic oscillators at resonance, Ann. Mat. Pura Appl. 82 (1969), 49–68.

[71] J. Leray and J. Schauder: Topologie et équations fonctionnelles. Ann. Sci. Ecole Norm. Sup. 3 No 51 (1934), 45–78.

[72] N. Lloyd: Degree theory. Cambridge University Press, Cambridge (1978).

[73] J. Mawhin: Landesman-Lazer conditions for boundary value problems: A nonlinear version of resonance. Bol. de la Sociedad Española de Mat. Aplicada 16 (2000), 45–65.

[74] J. Mawhin: Topological degree methods in nonlinear boundary value problems, NSF-CBMS Regional Conference in Mathematics no. 40, American Mathematical Society, Providence, RI (1979).

[75] J. Mawhin: Boundary value problems for nonlinear second-order vector differential equations, J. Differential Equations 16 (1974), 257–269.

[76] J. Mawhin: The Bernstein-Nagumo problem and two-point boundary value problems for ordinary differential equations. Qualitative Theory of Differential Equations, Farkas ed., Budapest (1981), 709–740.

[77] J. Mawhin and M. Willem: Critical point theory and Hamiltonian systems. Springer, New York (1989).

[78] L. Nirenberg: Generalized degree and nonlinear problems, Contributions to nonlinear functional analysis, Ed. E. H. Zarantonello, Academic Press New York (1971), 1–9.

[79] J. Mawhin: The forced pendulum: A paradigm for nonlinear analysis and dynamical systems. Expo. Math., 6 (1988), 271–87.

[80] J. Mawhin: Boundary value problems for nonlinear ordinary differential equations: from successive approximations to topology. Development of mathematics 1900–1950 (Luxembourg, 1992), 443–477, Birkhäuser, Basel, 1994.

[81] J. Mawhin: Topological Degree Methods in Nonlinear Boundary Value Problems. Regional Conference Series in Mathematics 40, Amer. Math. Soc., Providence, RI (1979).

[82] J. Mawhin: An extension of a theorem of A. C. Lazer on forced nonlinear oscillations, J. Math. Anal. Appl. 40 (1972), 20–29.

[83] J. Mawhin: Continuation theorems and periodic solutions of ordinary differential equations. Topological Methods in Differential Equations and Inclusions. NATO Science Series C: Mathematical and Physical Sciences. Editors A. Granas, M. Frigon, G. Sabidussi. Springer Netherlands (1995), 291–375.

[84] J. Mawhin: Leray–Schauder degree: a half century of extensions and applications. Topological Methods in Nonlinear Analysis 14 (1999), 195–228.

[85] J. Mawhin: Topological Fixed Point Theory and Nonlinear Differential Equations. Handbook of Topological Fixed Point Theory, R.F. Brown et al. eds., Springer, Berlin, 2005, 867–904.

[86] J. Mawhin and M. Willem: Multiple solutions of the periodic boundary value problem for some forced pendulum-type equations. J. Differential Equations 52 (1984), 264–287.

[87] J. Milnor: Topology from a Differential Viewpoint. University of Virginia Press (1965).

[88] J. Milnor: Analytic proofs of the "hairy ball theorem" and the Brouwer fixed-point theorem, Amer. Math. Monthly 85 (1978), no. 7, 521–524.

[89] M. Nagumo: Uber die differentialgleichung $y'' = f(t, y, y')$. Proc. Phys-Math. Soc. Japan 19 (1937), 861–866.

[90] M. Nagumo: On principally linear elliptic differential equations of the second order. Osaka Math. J. 6 (1954) 207–229.

[91] M. Nagumo: A theory of degree of mapping based on infinitesimal analysis. Amer. J. Math. 73 485–496 (1951).

[92] M. Nagumo: Degree of mapping in convex linear topological spaces. Amer. J. Math. 73 497–511 (1951).

[93] L. Nirenberg: Topics in Nonlinear Functional Analysis. Courant Lecture Notes - Sciences, New York (1974).

[94] R. Ortega: A counterexample for the damped pendulum equation, Acad. Roy. Belg. Bull. Cl. Sci. 73 (1987), 405–409.

[95] R. Ortega: Counting periodic solutions of the forced pendulum equation. Nonlinear Analysis 42 (2000) 1055–1062.

[96] R. Ortega and L. Sánchez: Periodic solutions of forced oscillators with several degrees of freedom, Bull. London Math. Soc. 34 (2002), 308–318.

[97] R. Ortega, E. Serra and M. Tarallo: Non-continuation of the periodic oscillations of a forced pendulum in the presence of friction. Proc. of Am. Math. Soc. 128, 9 (2000), 2659–2665.

[98] R. Ortega; M. Tarallo: Degenerate equations of pendulum-type. Commun. Contemp. Math. 2 (2000), 127–149.

[99] R. Palais: A simple proof of the Banach contraction principle. Journal of Fixed Point Theory and Applications 2 (2007), 221–223.

[100] E. Picard: Sur l'application des méthodes d'approximations succesives à l'étude de certaines équations différentielles ordinaires, J. Math. Pures Appl. 9 (1893), 217–271.

[101] C. Rogers: A Less Strange Version of Milnor's Proof of Brouwer's Fixed-Point Theorem. The American Mathematical Monthly, Vol. 87, 7 (1980), 525–52.

[102] D. Ruiz and J. R. Ward Jr.: Some notes on periodic systems with linear part at resonance. Discrete and Continuous Dynamical Systems 11 (2004), 337–350.

[103] A. Sard: The measure of critical values of differentiable maps. Bull. Amer. Math. Soc. 48 (1942) 883–890.

[104] K. Schmitt, R. Thompson: Nonlinear Analysis and Differential Equations. An Introduction. University of Utah (2004).

[105] G. Scorza-Dragoni: Il problema dei valori ai limiti studiato in grande per gli integrali di una equazione differenziale del secondo ordine, Giornale di Mat. (Battaglini) 69 (1931), 77–112.

[106] C. Severini: Sopra gli integrali delle equazione differenziali del secondo ordine con valori prestabiliti in due punti dati, Atti R. Acc. Torino 40 (1904–5), 1035–40.

[107] G. Teschl: Nonlinear Functional Analysis. Lecture notes, University of Vienna (2005).

[108] G. Whyburn, Topological Analysis, Princeton University Press, Princeton, NJ, (1964).

[109] E. Zeidler: Nonlinear Functional Analysis and Its Applications. Part 1: Fixed-Point Theorems. Springer-Verlag, Berlin, Heidelberg, New York, Tokyo (1985).

[110] G. Zhang and S. Cheng: Positive periodic solutions of nonautonomous functional differential equations depending on a parameter. Abstr. Appl. Anal. 7 (2002), 279–286.

Index